INTRODUCTION TO CHEMICAL PRINCIPLES

a laboratory approach

SUSAN A. WEINER

Department of Chemistry,
West Valley College, Saratoga, California

EDWARD I. PETERS

Department of Chemistry,
West Valley College, Saratoga, California

SAUNDERS GOLDEN SUNBURST SERIES

W. B. SAUNDERS COMPANY • Philadelphia • London • Toronto

W. B. Saunders Company: West Washington Square
Philadelphia, PA 19105

1 St. Anne's Road
Eastbourne, East Sussex BN21 3UN, England

1 Goldthorne Avenue
Toronto, Ontario M8Z 5T9, Canada

Introduction to Chemical Principles – A Laboratory Approach   ISBN   0-7216-9159-5

Last digit is the print number: 9   8   7   6   5

PREFACE

This laboratory manual is addressed to the students enrolled in a one-semester introductory chemistry course and to the teachers responsible for such a course. "Introductory" means many things; in this context it refers either to the class having the specific purpose of preparing its students for a full year freshman-level general chemistry course to follow; or to the less demanding type of general chemistry course for non-science majors given in many colleges and universities.

While this manual is a most suitable companion for *Introduction to Chemical Principles,* a textbook by one of its authors, this is more because both books have been written for the same course rather than the result of a deliberate attempt to match the two presentations. The laboratory manual is quite compatible with other fine textbooks that have been written for either of the student groups mentioned above.

At the semester's end, the student finishing this laboratory program will have acquired some experience in the more common laboratory operations; he will have developed and improved his ability to observe and propose tentative explanations for what he observes; and he will have made laboratory measurements and used them in analyzing his data. These experiences will cover a broad range of chemical topics, theories and principles typical of those normally considered in an introductory course. The student also will have gained some insight into the underlying role the laboratory plays in all chemical investigation. Finally, he will be well prepared for the more rigorous demands that will be made upon him in the general chemistry program.

The order in which the experiments appear is one of several that might be called typical, but it is not restrictive in any way. Aside from two titration experiments, each laboratory exercise is independent of all others. The instructor may therefore select the experiments in any order he finds most convenient for his purposes. With the exception of Experiment 23, which requires the weighing of a dried product at the beginning of the following laboratory session, all experiments are written so they may be completed comfortably in a three hour laboratory period, even if a large portion of the first hour is used for pre-laboratory or other discussion purposes.

In every introductory chemistry course, the period during which the lecture section considers atomic structure and chemical bonding is difficult to support with meaningful experiments. It has been our purpose, therefore, to supply enough early experiments to furnish an ample backlog based on the first few topics of the course to bridge over this period of little correlation between lecture and laboratory. Experiments 1 to 4 may be done with no matching lecture or text coverage, and experiments through number 8 are in order if the instructor considers formulas and equations quantitatively prior to exploring the details of atomic structure.

Anyone who has taught a chemistry laboratory section is well aware of the better results that are attained in less time by the student who has studied the experiment *before* the laboratory period begins. He is probably equally aware of the *disinclination,* shall we say, of most students to avail themselves of this advantage. To encourage preparation for the laboratory sessions, we have followed the highly successful example of Slowinski, Masterton and Wolsey in their *Chemical Principles in the Laboratory* by providing Advance Study Assignments for each experiment. These are to be completed by the student prior to the laboratory period, and turned in at its very beginning. These assignments consist of a few questions that are readily answered after a preliminary reading of the experiment.

A second aid to the student takes the form of performance goals. One or more sentences at the beginning of the experiment identify precisely what capability the student is to acquire as a result of completing the experiment. With the goal clearly established in the student's mind, progress toward it is greatly enhanced.

We have also included an aid to the instructor, the benefits of which will reach the students as well. Keenly aware of the staggering number of hours that are required to evaluate laboratory reports, we have kept our report sheets to a single page in most experiments and have restricted the questions to those points that are clearly essential to the purpose of the experiment. Hopefully, this will make grading less time-consuming and more practical.

The names that appear on the cover of a book rarely list all of the people who have contributed to its preparation. This book is no exception. Among the behind-the-scenes people who share whatever credit this book may earn are Mrs. Aileen Martin, who coordinated and arranged much of the physical preparation of the manuscript, and Mrs. Pat Baker, who did much of the typing. Our colleagues at West Valley College, Mrs. Betty Michelozzi and Dr. Thomas Hall, were most helpful in their willingness to try some of these experiments in their classes and offer valuable suggestions on how they might be improved. Mrs. Dorothy Steele deserves particular credit for being the first with the courage to use the preliminary manuscript for an entire course. Her observations and recommendations have aided us greatly. Finally, there are the students who rose to the challenge of new and untried material and shared their impressions with us. To all of these otherwise unacknowledged friends and aides, the authors express their sincere thanks.

SUSAN A. WEINER

EDWARD I. PETERS

CONTENTS

Laboratory Rules

SAFETY IN THE LABORATORY

A chemistry laboratory can be, and should be, a safe place to work in. Accidents can be prevented by observing safety rules, using proper judgment and carefully following the directions outlined in the experiments. In addition to the rules below, specific instructions are given in each experiment to eliminate potential hazards. *Do not* deviate from the procedures given in the experiments unless you are specifically instructed to do so. **THERE IS NO SUBSTITUTE FOR SAFETY.** Learn and observe these safety rules.

1. Do not eat, drink or smoke in the laboratory.
2. Eye protection (goggles, safety glasses) must be worn by all students when working in the laboratory.
3. Do not taste any chemical.
4. Never perform any unauthorized experiment.
5. Do not work in the laboratory unless an instructor is present.
6. Never point the open end of a test tube at yourself or at another person.
7. When inserting a glass tube, rod or thermometer into a rubber tube or stopper, be sure to protect your hands by holding the unit with gloves or layers of paper towel. Lubricating the glass surface is helpful.
8. Hot glassware looks no different from cold glassware. Place a reminder near it to warn others that the object is hot.
9. All experiments or operations which produce poisonous or noxious fumes **MUST** be performed in a fume hood.
10. Purses, sweaters, lunch bags and extra books should be stored in designated areas but not in the laboratory working area.
11. When diluting acids, always add the acid to water, never the other way around.
12. When disposing of liquid chemicals in a sink, flush with large amounts of cold water.
13. Do not dispose of solid chemicals in the sinks.
14. For organic solvent disposal, consult your instructor.
15. Most organic solvents are flammable. Keep these liquids away from an open flame.
16. If you want to smell a substance, do not hold it directly to your nose; instead, hold the container a few inches away and use your

1

hand to direct the vapors toward you.

17. Long hair should be tied back or pinned up; it tends to fall into chemicals or flames.
18. No bare feet are allowed in the laboratory.
19. Before leaving the laboratory, wipe the desk tops and wash your hands with soap and water.

Generally, laboratories are equipped with fire extinguishers, fire blankets, emergency showers, and so forth. Make sure you learn where these devices are and how to operate them.

PREVENTING CONTAMINATION OF CHEMICALS

In order to conduct experiments successfully in a laboratory, you must avoid contaminating the chemical reagents you use. The following procedures will help minimize the possibility of contamination.

(1) When a portion of a chemical is removed from its original container (regardless of whether it is a solid or a liquid), *do not* return the excess to the stock bottle. Dispose of the unused portion as you are directed by your instructor.

(2) After washing glassware, always use a final rinse of distilled (or deionized) water.

(3) Never remove liquid directly from a stock bottle. Pour a small portion into a clean, dry beaker and then use an eye dropper or pipet.

(4) Use separate spatulas to remove individual chemicals.

(5) Avoid handling more than one reagent bottle at a time; otherwise, you might interchange their stoppers by mistake.

(6) Do not lay reagent bottle tops or stoppers on desk tops.

(7) Never weigh a chemical directly on a balance pan. Use a preweighed container.

(8) Place covers on stock bottles when you are finished with them.

Properties and Changes of Matter

PERFORMANCE GOALS

1-1 Determine experimentally the solubility of a pure substance in a given liquid; or, in the case of two liquids, determine their miscibility.

1-2 Determine experimentally which of two immiscible liquids is more dense.

1-3 Determine whether or not a chemical reaction occurs when you combine two solutions, and state the criterion for your decision.

CHEMICAL OVERVIEW

All material things that compose our universe are referred to as *matter*. Matter is commonly defined as that which has mass and occupies space. In this experiment, we shall examine some of the characteristics of matter and introduce some of the language of science in which these characteristics are described.

A **pure substance** is a sample of matter that has identical properties throughout, and a definite, fixed composition. Matter composed of two or more pure substances is known as a **mixture**. Mixtures do not have a fixed composition; mixtures of the same two substances may vary widely in the proportion of one to the other. Mixtures are of two types, (a) **homogeneous** and (b) **heterogeneous**. Homogeneous mixtures have identical properties throughout, and the components are not visually distinguishable (i.e., you cannot see the sugar in a solution of sugar in water). By contrast, the components of a heterogeneous mixture are readily seen. Peanut brittle is a typical example of a heterogeneous mixture; its components are not uniformly distributed and are easily distinguishable.

Each pure substance can be characterized by a set of distinct properties. **Physical properties** are those inherent characteristics of a substance which can be observed without changing its composition. Common physical properties are taste, color, odor, melting and boiling points, solubility and density. **Chemical properties** describe the behavior of a substance when it reacts with other substances: its ability to burn, to react with water or even to decompose into two or more other pure substances.

3

Matter can undergo two types of changes, physical and chemical. **Physical changes** do not cause a change in composition, only in appearance. For example, when copper is melted, only a change of state occurs; no new substance is formed. In a **chemical change,** substances are converted into products having properties and compositions that are entirely different from those of the starting materials. Wood, for example, undergoes a chemical change when it burns (combines with oxygen).

When two liquids are mixed, they may form a homogeneous mixture. In this case, the liquids are **miscible.** Some liquids are miscible in all proportions, while others have a limited range of miscibility. If the two liquids are not miscible, i.e., **immiscible,** two distinct layers will form when they are poured together. The liquid having the lower density will "float" on top of the other.

When a solid is added to and dissolves in a liquid, it is **soluble** in the liquid. The mixture formed is called a **solution.** If the solid is **insoluble,** it will form a cloudy mixture from which the solid particles will settle slowly to the bottom.

When two solutions are combined, there may be a chemical reaction that yields a solid product. This solid is called a **precipitate**. It can be separated from the liquid by passing the mixture through a filter paper **(filtration)**. The solid particles will be retained by the paper, while the liquid will pass through and can be collected below. In other cases, the reaction may produce a **gaseous product** (effervescence) or soluble products which remain in solution. The appearance of a **color different from those of the starting components** is evidence of a reaction yielding a dissolved product. In many cases, no reaction occurs when two solutions are brought together.

PROCEDURE

1. MIXING LIQUIDS

a. Pour about one milliliter of carbon tetrachloride into a medium-size test tube. Add 10 drops of water with an eye dropper and gently shake the test tube. Are the two liquids miscible?

If the two liquids are *not* miscible, continue as follows: Reverse the sequence of placing the liquids into the test tube; i.e., start with one milliliter of water and add 10 drops of carbon tetrachloride. Record whether the result is the same or different from the first mixing. Determine from your observation which liquid is more dense.

b. Repeat the above procedure with methyl alcohol and water. Report your observations.

c. Pour about one milliliter of water into a test tube and add 10 drops of benzene. Gently shake the test tube and observe whether the two liquids are miscible. From your results in step 1(a) and this step, can you determine if benzene is less dense or more dense than carbon tetrachloride? Report and explain your answer.

d. Pour about one milliliter of carbon tetrachloride into a test tube and add 10 drops of benzene. Are the two liquids miscible? By mixing these two liquids, could you determine which has the greater density? Why or why not?

Note: Keep benzene and carbon tetrachloride bottles in a fume hood. Avoid breathing vapors.

2. DISSOLVING A SOLID IN A LIQUID

a. Place a small amount (about ¼" on the tip of a spatula) of barium chloride in a medium-size test tube. Add about one milliliter of distilled water and shake the test tube gently. Record your observations and save the solution for further use.

b. Repeat the procedure with sodium sulfate. Record your observations and save the solution.

c. Mix the contents of the test tubes in steps 2(a) and 2(b). Record your observations. Explain your results.

d. Repeat the procedure in step 2(a) with barium sulfate. Record your observations.

3. MIXING SOLUTIONS

a. Following the same procedure as in Part 2, prepare and mix solutions of sodium chloride and barium nitrate. Record your observations. Has a chemical reaction taken place? Explain.

b. Repeat the above procedure with solutions of iron (III) chloride and potassium thiocyanate. Record evidence of a chemical reaction, if any.

c. Prepare a solution of sodium carbonate. Add 2 to 3 drops of 3 M hydrochloric acid. Record evidence of a chemical reaction, if any.

d. Prepare a solution of copper (II) sulfate and add to it concentrated ammonia (often labeled NH_4OH), a drop at a time. Record evidence of a chemical reaction, if any.

e. Prepare and mix solutions of barium chloride and potassium chromate. Record evidence of a chemical reaction, if any.

**EXPERIMENT 1
REPORT SHEET**

Name _____

Date _____ Section _____

1. Mixing Liquids

a. Water plus carbon tetrachloride: Are they miscible (____) or

    immiscible (____) (check one)?

    Top layer, if immiscible: _____ .

    Which liquid is more dense? _____ .

b. Water and methyl alcohol: Miscible (____) or immiscible (____)?

    Top layer, if immiscible: _____ .

    Which liquid is more dense? _____ .

c. Water and benzene: Miscible (____) or immiscible (____)?

    Top layer, if immiscible: _____ .

    Which liquid is more dense? _____ .

    Can you determine if benzene is less dense or more dense than carbon
    tetrachloride? Explain.

d. Carbon tetrachloride plus benzene: Miscible (____) or immiscible

    (____)?

    Top layer, if immiscible: _____ .

    Can you determine from this step alone if benzene is less dense or
    more dense than carbon tetrachloride? Explain.

EXPERIMENT 1
REPORT SHEET

Name _____

Date _____ Section _____

Page 2

2. Dissolving a Solid in a Liquid

a. Barium chloride: Soluble (____) or insoluble (____)?

b. Sodium sulfate: Soluble (____) or insoluble (____)?

c. Mixing barium chloride and sodium sulfate solutions:

Observation: _____ .

Explanation: _____ .

d. Barium sulfate: Soluble (____) or insoluble (____)?

3. Mixing Solutions

a. Sodium chloride plus barium nitrate: Reaction (____) or no reaction

(____)? Evidence:

b. Iron (III) chloride plus potassium thiocyanate: Reaction (____) or no

reaction (____)? Evidence:

c. Sodium carbonate plus hydrochloric acid: Reaction (____) or no

reaction (____)? Evidence:

d. Copper (II) sulfate plus ammonia: Reaction (____) or no reaction

(____)? Evidence:

e. Barium chloride plus potassium chromate: Reaction (____) or no

reaction (____)? Evidence:

EXPERIMENT 1
ADVANCE STUDY ASSIGNMENT

Name _____

Date _____ Section _____

1. List five common physical properties of water.

2. Classify each of the following as a physical or chemical change:

    a. Melting ice _____

    b. Burning gasoline _____

    c. Dissolving sugar _____

    d. Copper forming copper (II) sulfide with sulfur _____

3. Identify three forms of evidence that a chemical reaction has occurred:

    a.

    b.

    c.

Densities of Liquids and Solids

PERFORMANCE GOALS

2-1 Calculate the density of a liquid or solid from experimental data.

CHEMICAL OVERVIEW

One of the physical properties that characterize a substance is its **density**, or its **mass per unit volume**. Mathematically,

$$\text{Density} = \frac{\text{mass}}{\text{volume}} \qquad (2.1)$$

According to this equation, density is equal to the ratio of the mass of a sample of a substance to the volume it occupies. The density of a solid is normally expressed in grams per cubic centimeter (g/cc or g/cm^3), the density of liquids in grams per cubic centimeter or grams per milliliter (g/ml), and the density of gases in grams per liter (g/liter).

Determination of the density of a substance is based on the measurement of both the volume and mass of the same sample of the substance. Mass expresses the quantity of matter. It is determined by comparing the mass of a sample to a known mass through the process commonly called *weighing*. The volume of a liquid can be measured with a calibrated container, such as a graduated cylinder. The volume of a solid can be measured directly only if the solid has a regular geometrical shape (i.e., a cube, a sphere or other shapes). Ordinary solid samples generally have irregular shapes and sizes, however. In such cases, the volume of the solid may be determined by measuring the volume of a liquid displaced when the solid sample is immersed in the liquid. The liquid should be chosen so that it (a) does not react with the solid, (b) does not dissolve the solid and (c) has a density lower than that of the sample.

In the first part of this experiment, you will be asked to determine experimentally the density of a known substance and then to calculate the

13

percent error in your determination. Percent error is defined by the following equation:

$$\text{Percent error} = \frac{\text{error}}{\text{accepted value}} \times 100 \qquad (2.2)$$

The "error" is the difference between the experimental value and the accepted value. Error is usually expressed as an *absolute value,* i.e., a numerical value only, disregarding any algebraic sign. Thus, Equation 2.2 becomes

$$\text{Percent error} = \frac{|\text{experimental value} - \text{accepted value}|}{\text{accepted value}} \times 100 \qquad (2.3)$$

where the vertical lines in the numerator designate the absolute value of the difference indicated.

PROCEDURE

1. DENSITY OF A LIQUID

Weigh a clean, dry 25 ml graduated cylinder to the nearest 0.01 grams on a centigram balance. Remove the cylinder from the balance and pour into it 12 to 15 ml of carbon tetrachloride. Wipe the outside dry and weigh it again. Estimate the volume of the liquid to the nearest 0.1 ml, always reading the bottom of the meniscus at eye level (see Figure 2-1). Record your measurements on your report sheet. The accepted value of the density of carbon tetrachloride is 1.59 g/ml.

In the same manner, determine the density of an unknown liquid.

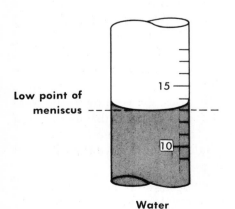

Low point of
meniscus

Water

Figure 2-1. Determining the lowest point of a meniscus in volume reading.

2. DENSITY OF A SOLID

Pour 10 to 12 ml of any liquid into the same preweighed graduated cylinder used in Part 1, and weigh the cylinder and liquid to the nearest 0.01 grams on a centigram balance. (You may use the unknown non-aqueous liquid in Part 1, thereby using the second run to check the density of your unknown liquid.) Record the volume of the liquid to the nearest 0.1 ml. Add an unknown solid until you note a volume increase of 8 to 10 ml. Tap the sides of the cylinder to dislodge any air bubbles. Again read the volume — this time the volume of the liquid plus the solid — and weigh. The volume increase is equal to the volume of the solid added. Record your measurements on the report sheet.

CALCULATIONS

1. DENSITY OF A LIQUID

The mass of the liquid is obtained by difference — by subtracting the mass of the empty cylinder from the mass of the cylinder plus the liquid. Liquid density is found by dividing mass in grams by volume in milliliters, as indicated in Equation 2.1. Percent error may be calculated by substituting into Equation 2.3.

Note: Keep carbon tetrachloride bottle in fume hood, and avoid breathing vapors.

2. DENSITY OF A SOLID

Both the volume and the mass of the sample solid are obtained by difference. The density is calculated by substitution into Equation 2.1.

EXPERIMENT 2
REPORT SHEET

Name _____

Date _____ Section _____

1. Density of a Liquid

Unknown Liquid No. _____

	Carbon Tetrachloride	*Unknown*
Mass of cylinder + liquid	_____ g	_____ g
Mass of cylinder	_____ g	_____ g
Mass of liquid	_____ g	_____ g
Volume of liquid	_____ ml	_____ ml

Density of liquid
  (Show calculation setups.)

_____ g/ml   _____ g/ml

% Error for carbon tetrachloride
  (Show calculation setup.)

_____ % error

2. Density of a Solid

Unknown Solid No. _____

	Trial 1	Trial 2
Mass of cylinder + liquid + solid	_____ g	_____ g
Mass of cylinder + liquid	_____ g	_____ g
Mass of solid	_____ g	_____ g
Volume of liquid and solid	_____ ml	_____ ml
Volume of liquid	_____ ml	_____ ml
Volume of solid	_____ cc	_____ cc

Density of solid
(Show calculation setups.)

_____ g/cc   _____ g/cc

EXPERIMENT 2

ADVANCE STUDY ASSIGNMENT

Name _____

Date _____ Section _____

1. A piece of metal measures 2.00 cm X 4.00 cm X 10.0 cm and weighs 520 grams. Calculate its density.

2. A liquid has a density of 0.65 g/ml. Calculate the mass of 250 ml of that liquid.

3. How many milliliters of solution should you measure to obtain 35.0 grams if its density is 1.06 g/ml?

Separation of Cations by Paper Chromatography

PERFORMANCE GOALS

3-1 Separate a mixture of cations by paper chromatography and calculate their R_F values.

3-2 Analyze an unknown mixture of cations by paper chromatography.

CHEMICAL OVERVIEW

Chromatography, which means "the graphing of colors," obtains its name from the early experiments of Tswett, who, in 1906, succeeded in separating a mixture of colored pigments obtained from leaves. A solvent mixture, carrying the pigments, was allowed to pass through a glass column packed with chalk. At the end of the experiment, the pigments were separated in colored bands at various distances from the starting level. This method is now known as column chromatography.

Chromatography may now be applied to colorless compounds and to ions. Paper chromatography is a more recent and much faster separation technique than column chromatography. It may be used for the separation of substances by a solvent moving on sheets or strips of filter paper. The filter paper is referred to as the **stationary phase**, or **adsorbent**. The mixture of solvents used to carry the substances along the paper is called the **mobile phase**, or **solvent system**.

In practice, a sample of the solution containing the substances to be separated is dried on the paper. The edge of the paper is dipped into the solvent system so that the separation sample is slightly above the liquid surface. As the solvent begins to soak the paper, rising by capillary action, it transports the sample mixture upward. Each component of the mixture being separated is held back by the stationary phase to a different extent. Also, each component has a different solubility in the mobile phase and therefore moves forward at a different speed. A combination of these effects causes each component of the mixture to progress at a different rate, resulting in separation.

In a given solvent system, using the same adsorbent at a fixed temperature, each substance can be characterized by a constant R_F value. By definition.

$$R_F = \frac{\text{Distance from origin to center of spot}}{\text{Distance from origin to solvent front}} \qquad (3.1)$$

where the *origin* is the point at which the sample was originally placed on the paper and the *solvent front* is the line representing the most advanced penetration of the paper by the solvent system. The R_F value is a characteristic property of a species, just as the melting point is a characteristic property of a compound.

In this experiment, we shall separate a mixture of iron (III), copper (II) and nickel (II) ions, Fe^{3+}, Cu^{2+} and Ni^{2+}, respectively. Fe^{3+} ions produce a rust color on wet paper and are therefore directly visible. The presence of Cu^{2+} ions will be indicated by the deep blue complex formed by their reaction with ammonia. Ni^{2+} ions will be identified by the red complex they form with an organic reagent, dimethylglyoxime.

PROCEDURE

Working in a hood, prepare the following solvent system, using a graduated cylinder: acetone, 19 ml; concentrated hydrochloric acid, 4 ml; water, 2 ml.

CAUTION: ACETONE IS EXTREMELY FLAMMABLE. VAPORS CAN IGNITE EVEN WHEN THE LIQUID IS A CONSIDERABLE DISTANCE FROM AN OPEN FLAME. THE FUMES OF HYDROCHLORIC ACID ARE HARMFUL, AND IT HAS AN OBNOXIOUS, SUFFOCATING ODOR. CONCENTRATED HYDROCHLORIC ACID IS ALSO HARMFUL TO THE SKIN.

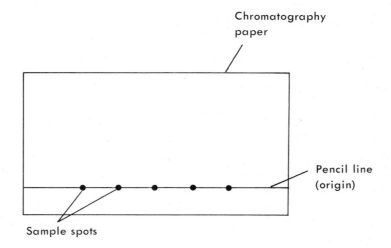

Figure 3-1. Preparing chromatography paper.

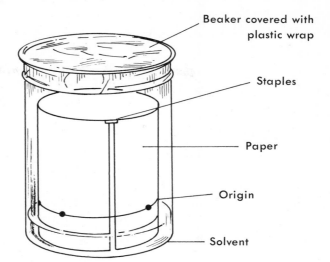

Figure 3-2. Development of chromatogram.

Pour the solvent mixture into an 800 ml beaker and cover the beaker tightly with a plastic film (e.g., Saran Wrap). This procedure allows the atmosphere within the beaker to become saturated with solvent vapor and helps to give a better chromatographic separation.

Obtain a piece of chromatography paper measuring 24.5 cm by 12.5 cm. Draw a line in pencil (do NOT use ink) about one cm away from a long edge of the paper. This line will indicate the origin (see Figure 3-1).

Using capillary tubes, transfer a small drop of each of the five solutions listed below to the penciled line. Apply the spots evenly over the line, leaving a margin of about 3 cm from either short edge of the paper strip and using a separate, clean capillary tube for each solution. With a pencil, identify each spot by writing directly on the filter paper. The solutions are:
   a.   A solution containing a soluble Fe^{3+} compound.
   b.   A solution containing a soluble Cu^{2+} compound.
   c.   A solution containing a soluble Ni^{2+} compound.
   d.   A solution containing a mixture of all three ions.
   e.   A solution containing one, two or all three of the ions.

Allow the paper to dry; an air blower or heat lamp may be used to speed up drying. Form the paper into a cylinder without overlaping the edges. Fasten the paper with staples and, taking care that the origin line is above the solvent, place it in the beaker, as in Figure 3-2. Replace the plastic film cover over the beaker and wait until the solvent has moved up the paper about 6 to 7 cm from the origin. Remove the paper quickly, mark the solvent front position with a pencil.

Find any spots that appear colored and circle them in pencil. Working in a fume hood, pour a few milliliters of concentrated ammonia (often labeled NH_4OH) into an evaporating dish, and hold the chromatogram over the dish. Circle any new spots that become visible on contact with the gaseous ammonia. While still moist with ammonia, run the paper

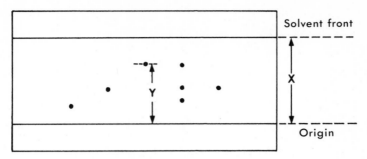

Figure 3-3. Developed chromatogram.

through a few milliliters of dimethylglyoxime in a shallow dish (an evaporating dish is satisfactory; a petri dish is better). Dry the paper and circle any new spots.

RESULTS AND CALCULATIONS

Observe and record on the report sheet the colors of spots produced by the Fe^{3+}, Cu^{2+} and Ni^{2+} ions in each of the chromatograms of solutions (a) through (d).

Measure and record in millimeters the distance between the origin and the solvent front (x in Figure 3-3). Next, measure and record the distance between the origin and the center of each spot in the chromatogram for all solutions (y in Figure 3-3). Calculate the R_F value for each ion, using Equation 3.1. Record the result in the report sheet.

Now determine both the color and R_F value of each spot resulting from your unknown solution (e), and compare them with the results from your known solutions. Record these in your report sheet, and indicate which ions were present in your unknown.

EXPERIMENT 3
REPORT SHEET

Name _____

Date _____   Section _____

Solution	Color			Distance from Origin (mm)			R_F Value		
	Fe^{3+}	Cu^{2+}	Ni^{2+}	Fe^{3+}	Cu^{2+}	Ni^{2+}	Fe^{3+}	Cu^{2+}	Ni^{2+}
(a)		—	—		—	—		—	—
(b)	—		—	—		—	—		—
(c)	—	—		—	—		—	—	
(d)									
(e)									

Unknown No. _____

Distance of solvent front from origin: _____ mm.

Ions present in and observed R_F values from unknown:

EXPERIMENT 3   Name _____

ADVANCE STUDY ASSIGNMENT

Date _____ Section _____

1. Why should you use pencil rather than ink for markings on the chromatography paper?

2. What error might be caused by applying too large a drop of liquid on the origin line?

3. Why should you not overlap the edges of the chromatogram when it is being developed?

4. Why should you avoid moving the beaker while the chromatogram is being developed?

Calibration of a Thermometer

PERFORMANCE GOALS

4-1 Calibrate a thermometer by measuring melting or freezing points of pure substances.

4-2 Construct a calibration curve for a thermometer.

CHEMICAL OVERVIEW

More than most laboratory devices, thermometers need to be calibrated if the measured temperatures are to be considered reliable. The inaccuracy of a thermometer may be due to several causes, including irregularity in the capillary tube and thermal expansion and contraction during temperature cycling. Thermometers for precision laboratory work may be calibrated against a very highly accurate thermometer available from the National Bureau of Standards. Alternately, a thermometer can be calibrated experimentally by matching its readings with the known melting or freezing points of pure substances. The second method will be used in this experiment.

The melting point (or freezing point), a physical property of pure substances, is the temperature at which a substance changes between the solid and the liquid states. The melting behavior of a compound is often a simple test of its purity. An impure compound melts over a range of temperatures that are lower than the melting point of the pure compound.

In this experiment, calibration will be carried out by checking the accuracy of your thermometer at the following temperatures:

1. Melting point of ice, 0.0°C.
2. Freezing point of glacial acetic acid, 16.7°C.
3. Melting point of p-dichlorobenzene, 53.2°C.
4. Melting point of naphthalene, 80.2°C.
5. Melting point of α-naphthol, 93.4°C.

Some thermometers are designed for total immersion in the temperature zone being measured; others have an immersion line etched on them. In both cases, if you use the instrument as intended, you will probably minimize temperature errors. For best results, the calibration of either type of thermometer will be more accurate if it is carried out under the conditions of its intended use. Since most work with the thermometer in the general chemistry laboratory is performed when only the lower portion of the thermometer is immersed in the test zone, it is advisable to calibrate your thermometer for partial immersion. If the thermometer has no immersion line etched on it, choose a fixed position on the thermometer about 6 to 7 cm above the bulb and always immerse the thermometer to that depth.

PROCEDURE

1. MELTING POINT OF ICE

In a 250 ml beaker place more than enough clean crushed ice to cover the immersion line of the thermometer and add just enough distilled water to cover the ice. Place the thermometer in the slush to the immersion mark and stir the mixture with a stirring rod. When the level of the

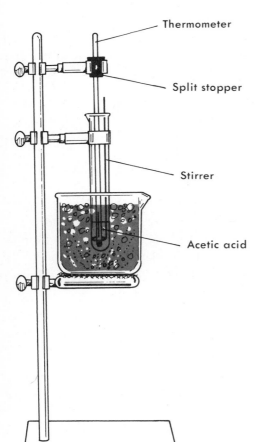

Thermometer

Split stopper

Stirrer

Acetic acid

Figure 4-1. Apparatus for freezing point determination.

mercury column remains constant for at least 1 minute, record the thermometer reading to the nearest 0.1°C. Enter your reading in the report sheet.

2. FREEZING POINT OF GLACIAL ACETIC ACID

Pour 15 ml of glacial acetic acid into a large, clean, dry test tube. Using a ring stand and a clamp, suspend the test tube in the beaker of ice used in Step 1 (see Figure 4-1). Be sure the level of the acetic acid in the test tube is below the level of slush in the beaker. Place a split stopper around the dry thermometer and, holding it with a second clamp, lower it into the test tube so that the immersion line is at the surface of the liquid. Loop a circular wire stirrer around the thermometer in the test tube, and by raising and lowering the wire stir the acetic acid. Observe the temperature change and record the temperature to 0.1°C when the first crystals appear and the temperature remains constant.

3. MELTING POINT OF p-DICHLOROBENZENE

Obtain a melting point capillary. Crush a small amount of p-dichlorobenzene on a watch glass with a spatula, and scrape the powder into a mound. Fill the capillary by pushing its open end into the mound and then compact the powder by allowing the vertically held capillary to drop onto a hard surface (such as a table top). Do not press too much powder into the capillary tube during each step, because the powder will pack near the opening and then cannot be removed to the bottom of the tube. Add material until you have a tightly packed column about three-fourths of the length of the thermometer bulb.

Fasten the capillary to the thermometer by means of a rubber band or a thin slice of rubber tubing, making sure the packed part of the capillary is level with the thermometer bulb. Fill a 250 ml beaker with water and assemble the apparatus shown in Figure 4-2. The water level should be below the open end of the capillary tube and at the immersion level of the thermometer. Heat the water slowly with constant stirring and carefully watch the capillary tube. Read the thermometer at the first appearance of a liquid phase and again when the last of the solid disappears. Record these temperatures. For the purpose of calculating the temperature correction, take the midpoint of the temperature range and compare it to the known melting point.

4. MELTING POINT OF NAPHTHALENE

Using a fresh capillary for each sample, determine and record the melting point of naphthalene by the same method described in Step 3.

5. MELTING POINT OF α- NAPHTOL

Repeat the procedure for Step 3 using α-naphthol in a clean capillary tube.

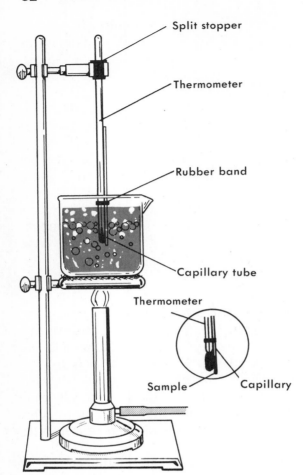

Split stopper

Thermometer

Rubber band

Capillary tube

Thermometer

Sample   Capillary

Figure 4-2. Apparatus for determination of melting point.

NOTE: It is possible to combine steps 3 through 5 by measuring the melting points of the substances in a continuous fashion. To do so, prepare three capillary tubes, one for each compound listed in Steps 3, 4 and 5. Fasten all three capillaries to the thermometer at the same time. Proceed as described in Step 3, heating the water slowly. Observe the melting ranges of the compounds in the order of increasing temperatures. Record the temperatures and calculate the thermometer correction for each case.

CALCULATIONS

In steps 3 to 5, record the midpoint of the temperature ranges as the melting point. For all steps, calculate the temperature correction.* List the correction as positive if it must be added to the observed reading to get the known, but negative if the correction must be subtracted.

———————————

*Mathematically: Correction = True value – Observed value.

(Example: Your thermometer is used to measure the freezing point of benzene, and the range is from 5.3 to 5.5°C. The known freezing point is 5.5°C. Correction: 5.5°C – 5.4°C = +0.1°C.) Prepare a correction curve on the paper provided by plotting the corrections against the temperature reading. Draw a smooth curve through the points. Submit this curve with your report sheet.

EXPERIMENT 4
REPORT SHEET

Name _____

Date _____ Section _____

Measurement	Observed Melting Range	Observed Melting (Freezing) Point	Known Melting (Freezing) Point	Correction (Include Sign)
Melting of Ice	—			
Freezing of Acetic Acid	—			
Melting of p-dichlorobenzene				
Melting of Naphthalene				
Melting of α-naphthol				

Immersion Line: _____ (cm from bottom of bulb).

Name _____

Date _____ Section _____

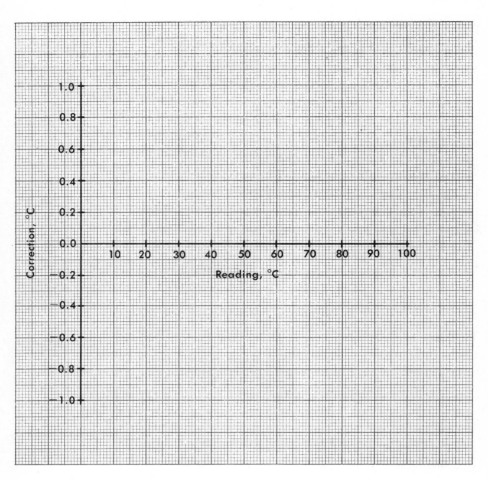

Calibration Curve of a Thermometer

EXPERIMENT 4

Name _____

ADVANCE STUDY ASSIGNMENT

Date _____ Section _____

1. In the calibration procedure, why must we use pure substances?

2. Why is it always important to immerse the thermometer to the same depth?

3. What is the difference between the melting point of ice and the freezing point of water?

Empirical Formula of a Compound

PERFORMANCE GOALS

5-1 Prepare a compound from a weighed quantity of metal.

5-2 Determine the mass of the compound, and from it calculate the mass of the nonmetal.

5-3 From the masses of the elements in the compound, determine its empirical formula.

CHEMICAL OVERVIEW

Chemical compounds are composed of atoms of various elements. The atoms are held together by chemical bonds. It has been shown experimentally that the ratio of moles of the constituent elements in the compound is nearly always a ratio of small, whole numbers. The few exceptions are known as nonstoichiometric compounds. The formula containing the lowest possible ratio is known as the *empirical formula*. At times it may be the same as the molecular formula; often, however, the molecular formula is an integral multiple of the simplest, empirical formula. For example, the empirical formula of the compound benzene (C_6H_6) is simply CH, indicating that the ratio of carbon atoms to hydrogen atoms in the compound is one to one.

To find the empirical formula of a compound, we must combine the elements forming the compound under conditions allowing us to determine the mass of each element. From these data, the moles of each element may be determined. By dividing these numbers by the smallest number of moles, we obtain quotients that are in a simple ratio of integers, or are readily converted to such a ratio.

SAMPLE CALCULATIONS

A strip of aluminum weighing 0.270 g is ignited, yielding an oxide that weighs 0.510 g. Calculate the empirical formula of the oxide.

STEP 1

Calculate the mass of the oxygen that combined with the aluminum.

$$\text{mass oxygen} = \text{mass product} - \text{mass aluminum}$$

$$\text{mass oxygen} = 0.510 \text{ g} - 0.270 \text{ g} = 0.240 \text{ g}$$

STEP 2

Calculate the moles of atoms of each element.

$$0.270 \text{ g Al} \times \frac{1 \text{ mole Al atom}}{27.0 \text{ g}} = 0.0100 \text{ moles Al atoms}$$

$$0.240 \text{ g O} \times \frac{1 \text{ mole O atom}}{16.0 \text{ g O}} = 0.0150 \text{ moles O atoms}$$

STEP 3

Obtain the ratio of atoms by dividing the moles of each by the smallest number of moles.

$$\text{O:} \frac{0.0150}{0.0100} = 1.50 \qquad \text{Al:} \frac{0.0100}{0.0100} = 1.00$$

The ratio is 1.50 atoms of oxygen to 1.00 atoms of aluminum. Change this ratio to a whole number ratio (multiplying by 2).

$$1.50 \times 2 = 3 \text{ O} \qquad 1.00 \times 2 = 2 \text{ Al}$$

Empirical formula: Al_2O_3

PROCEDURE

OPTION 1: A SULFIDE OF COPPER

Support a clean, dry porcelain crucible and its lid on a clay triangle (see Figure 5-1) and heat it slowly at first, then fairly strongly in the direct flame of a burner for about 4 to 5 minutes. Allow the crucible to cool to room temperature (never weigh any object while it is still hot!) and weigh it on a centigram balance.

Place a copper wire coil or medium shavings (about 1.5 to 2.0 grams) in the crucible and weigh them accurately. Cover the metal with 1 to 1.5 g

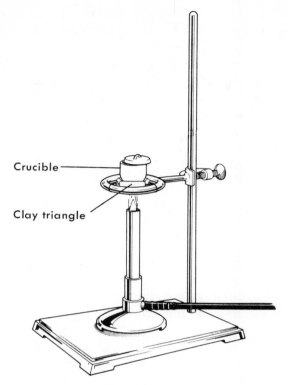

Crucible

Clay triangle

Figure 5-1.

of powdered sulfur and place a lid on the crucible. Start heating the crucible in a fume hood, slowly at first and then with a moderate flame, until the sulfur no longer burns around the lid. Finally, heat the crucible strongly for about 5 minutes more, making sure that no excess sulfur is present on the lid or on the sides of the crucible. Do not remove the lid until the contents of the crucible are cool, since air oxidation could possibly occur. When the crucible is cool, weigh again.

If there is a doubt that sufficient sulfur has been added or that the reaction has proceeded adequately, it is advisable to add a small amount of sulfur to the sample, heat it and weigh it again after is has cooled.

Knowing the mass of the copper sample and the mass of the sulfide, you can, by subtraction, calculate the mass of sulfur that reacted with the copper, and from that, the empirical formula of the compound can be calculated.

OPTION 2: AN OXIDE OF TIN

Place a clean porcelain crucible on a clay triangle (see Fig. 5-1) and heat it, first slowly, then with a hot flame, for about 4 to 5 minutes. Allow it to cool and weigh it on a centigram balance.

Select a piece of tin foil weighing about 1 to 1.5 grams and place it in the crucible in a loosely rolled ball. Weigh the crucible and the tin again to find the weight of tin by difference.

Under a fume hood, add concentrated HNO_3 drop by drop to the crucible until all of the tin has reacted and a damp white paste remains. Heat the paste cautiously with a mild flame, taking care not to cause spattering. After all of the liquid has evaporated, heat the crucible with a hot flame for 5 minutes. Cool it to room temperature and weigh it.

The compound formed from the tin during the reaction of tin is an oxide. By subtraction, you can calculate the amount of oxygen in the compound. From this information, calculate the empirical formula.

OPTION 3: MAGNESIUM OXIDE

Heat a clean porcelain crucible and its lid on a clay triangle (see Fig. 5–1), using a direct flame for 4 to 5 minutes. After allowing it to cool, weigh it on a *milligram* balance. Obtain a magnesium ribbon weighing 0.5 to 0.7 g, fold it into a loose ball (allowing for maximum surface) and place it in the crucible. Weigh the crucible and magnesium on a milligram balance.

Remove the lid and hold it near the crucible with a pair of tongs. Start heating the crucible, and as soon as the magnesium begins to burn, replace the lid. Continue the process, holding the escape of white smoke to a minimum (very finely divided magnesium oxide looks like smoke). When the contents of the crucible no longer burn, cock the lid on the crucible and heat it strongly for 5 minutes.

Since magnesium is an active metal, it combines with both oxygen and nitrogen when it is burned in air, forming the oxide MgO and the nitride Mg_3N_2. The nitride can be converted to the oxide according to:

$$Mg_3N_2 + 6H_2O \longrightarrow 3\,Mg(OH)_2 + 2\,NH_3 \qquad (5.1)$$

$$Mg(OH)_2 \longrightarrow MgO + H_2O \qquad (5.2)$$

Let the crucible cool, add about 10 drops of distilled water to it and then gently heat to vaporize the excess water. CAUTION: AVOID SPATTERING. Finish heating the crucible with a strong flame for 5 to 8 minutes. When the crucible is cool, weigh it again. Calculate the empirical formula from the weight of magnesium and the oxygen, the latter obtained by difference.

EXPERIMENT 5
REPORT SHEET

Name _____

Date _____   Section _____

	Sample 1	Sample 2
Mass of crucible (and lid, if used) and compound	_____ g	_____ g
Mass of crucible (and lid, if used) and metal	_____ g	_____ g
Mass of crucible (and lid, if used)	_____ g	_____ g
Mass of compound	_____ g	_____ g
Mass of metal	_____ g	_____ g
Mass of nonmetal element	_____ g	_____ g
Empirical formula (show calculations below)	_____	_____

EXPERIMENT 5
ADVANCE STUDY ASSIGNMENT

Name _____

Date _____   Section _____

1. 3.36 g of pure iron is allowed to react with oxygen to form the oxide. If the product weighs 4.64 g, calculate the formula of the compound.

2. You are to react metallic copper with powdered sulfur in the laboratory to form the sulfide of copper. How can you test whether the reaction has gone to completion?

3. Which of the following formulas is a correctly written empirical formula (circle one)?

   (a) $C_{1.5}H_4O_{0.5}$ (b) $(C_2H_4O)_2$ (c) $Mg_3P_2O_8$

Hydrates

PERFORMANCE GOALS

6-1 Calculate the percentage of the water of hydration from experimental data.

6-2 Calculate the formula of a hydrate of a known anhydrous salt from experimental data.

CHEMICAL OVERVIEW

Hydrates are chemical compounds that contain water as part of their crystal structure. This water is quite strongly bound and is present in a definite proportion relative to other constituents. It is referred to as **water of hydration.**

The stability of hydrates varies considerably. Some hydrates, on standing, spontaneously lose water to the atmosphere. Such compounds are described as being **efflorescent.** The opposite behavior is observed by other substances that tend to absorb water from the atmosphere. These compounds are referred to as **hygroscopic.** Some absorb so much water that they ultimately form a solution. These substances are said to be **deliquescent.**

Generally, water of hydration can be driven from hydrates by heating, leaving behind the **anhydrous** (without water) salt. This process may be accompanied by physical changes, such as a change of color. For example, $CuSO_4 \cdot 5H_2O$ is an intense blue shiny crystal which, upon heating, turns into very pale green-blue anhydrous $CuSO_4$.

In this experiment, you will be instructed to "heat to constant weight." In the context of the experiment, you want to heat the substance until *all* of the water is driven off. After a first heating, cooling and weighing, however, you cannot tell whether all water has been removed or if some still remains. You therefore repeat the heating, cooling and weighing procedure. If the same weight is reached after the second heating, you may assume that all water was removed the first time. If weight was lost in the second heating, you may be sure that all water was *not* removed in the first heating, and you are still unsure that it was all removed in the second heating. Another heating is therefore required. The heating, cooling and weighing sequence is repeated until two successive duplicate weighings are recorded.

PROCEDURE

1. BEHAVIOR OF HYDRATES

a. Place a few small crystals of $CoCl_2$ $6H_2O$ into a test tube and heat gently, holding the test tube tilted at an angle, and making sure that the top of the test tube is facing away from you and all others. Record observed changes. Allow the test tube to cool, and when it is cool, add a few drops of water. Hold the test tube against the back of your hand. Record your observations.

b. Efflorescence and Deliquescence. Place a few crystals of each of the compounds listed below on a watch glass and observe them every 20 minutes throughout the laboratory period. Record your periodic observations.

$Na_2SO_4 \cdot 10 H_2O$
$CaCl_2$
$MgSO_4 \cdot 7 H_2O$
$Na_2CO_3 \cdot 10 H_2O$

2. PERCENT WATER IN A HYDRATE; FORMULA OF A HYDRATE

Heat a clean porcelain crucible on a clay triangle over a direct flame for 5 minutes to drive off any surface moisture. Cool it and weigh it on a milligram balance. Heat it to constant weight in 3-minute heating cycles until duplicate weights (within 0.005 g) are reached. Place 1 to 1.5 grams of solid hydrate into the crucible and weigh the crucible again on a milligram balance. Heat the crucible and its contents, gently at first, then with a hot flame, for 10 minutes. Cool to room temperature and weigh. Heat it to constant weight in 3- to 4-minute heating cycles until duplicate weights are reached. Record all data in your report sheet.

CALCULATIONS

From the masses of (1) the crucible, (2) the crucible plus the anhydrous salt and (3) the crucible plus the hydrate, calculate by difference the weight of (1) the hydrate, (2) the anhydrous salt and (3) the water of hydration. Record these masses.

From the mass of water and the mass of the hydrate, calculate the percentage of hydration water in the hydrate. (Percentage may be determined by dividing the quantity of the component in question — water, in this case — by the total quantity to get the decimal fraction, and then multiplying by 100 to convert to percent.)

Your instructor will give you the formula of your anhydrous salt. Determine its molar weight and then calculate the moles of anhydrous salt in your sample. From the grams of water and its molar weight, compute the moles of water in the sample. From the ratio of these mole quantities, write the formula of the hydrate.

EXPERIMENT 6
REPORT SHEET

Name _____

Date _____ Section _____

1. Behavior of Hydrates

a. Changes observed with $CoCl_2 \cdot 6\,H_2O$:

b. Summarize your observations with each of the following and classify
each as efflorescent, hygroscopic or deliquescent:
$Na_2SO_4 \cdot 10\,H_2O$

$CaCl_2$

$MgSO_4 \cdot 7\,H_2O$

$Na_2CO_3 \cdot 10\,H_2O$

2. Percent Water in a Hydrate; Formula of a Hydrate

Unknown No. _____

Mass of crucible + hydrate   _____ g

Mass of crucible + anhydrous salt   _____ g

Mass of crucible   _____ g

Mass of hydrate   _____ g

Mass of anhydrous salt   _____ g

Mass of water of hydration   _____ g

Percent water of hydration   _____%
    (Show calculation setup)

Moles of anhydrous salt   _____ moles
    (Show calculation setup)

Moles of water   _____ moles
    (Show calculation setup)

Formula of the hydrate   _____

EXPERIMENT 6
ADVANCE STUDY ASSIGNMENT

Name _____

Date _____ Section _____

1. Calculate the percent water of hydration in $Na_2 CO_3 \cdot 10 \, H_2 O$.

2. A 1.500 gram sample of $Na_2 S_2 O_3 \cdot X \, H_2 O$ was heated until all of the water was removed. Calculate the formula of the hydrate if the residue after heating weighed 0.956 grams.

Calorimetry

PERFORMANCE GOALS

7-1 Determine a calorimeter constant.

7-2 Calculate the specific heat of an unknown solid element by measuring the heat exchanged in a calorimeter.

7-3 Using the Law of Dulong and Petit, calculate the approximate atomic weight of an unknown solid element.

CHEMICAL OVERVIEW

Heat, a form of energy, can be gained or lost by an object. When the object cools, it loses heat energy; when it is heated, it gains energy. The unit in which heat energy is measured is the *calorie (cal),* which is *the amount of heat required to raise the temperature of one gram of water by one degree Celsius.* The *kilocalorie (kcal)* is also frequently used; it is equal to 1000 calories.

The change in "heat content" experienced by an object as it passes from one known temperature to another, also called heat flow, can be calculated if the mass and *specific heat* of the object are known:

$$Q = (\text{mass}) \, (\text{specific heat}) \, (\Delta t), \qquad (7.1)$$

where Q is the heat flow in calories and Δt is the temperature change, or final temperature minus initial temperature. (The greek symbol Δ indicates change in a measured value, and always means final value minus initial value: $\Delta X = X_f - X_i$.) Specific heat, a property of a pure substance, expresses how many calories are required to raise the temperature of one gram of the substance by one degree Celsius. The units of specific heat are, from the definition, calories per gram degree, or cal/(g) ($^{\circ}$C). From the way the calorie is defined, the specific heat of water is 1.00 cal/(g) ($^{\circ}$C).

When a "hot" object comes into contact with a "cold" object — when an object at higher temperature comes into contact with an object at lower temperature — heat flows from the hot object to the cold object, until eventually they reach an intermediate temperature. Measurements of heat are made in well insulated devices called *calorimeters.* A perfect calorimeter is an isolated segment of the universe that allows no heat to

55

flow to or from its contents during an experiment. It follows from the law of conservation of energy that, in a perfect calorimeter,

$$\Sigma\, Q = 0 \qquad (7.2)$$

where $\Sigma\, Q$ is the sum of all the individual changes in heat content within the calorimeter — including the calorimeter itself.

Unfortunately there are no perfect calorimeters. While the calorimeter may reduce the heat transferred to or from the surroundings to a negligible amount, there is no way the calorimeter itself can be kept out of contact with its contents. Therefore, some heat will flow between the contents and the calorimeter. The calorimeter constant, K_c, is a measure of the number of calories transferred between the calorimeter and its contents per degree change *inside* the calorimeter. Expressed as an equation,

$$K_c = \frac{\text{calories transferred}}{\text{degrees of temperature change}} = \frac{Q_c}{\Delta t} \qquad (7.3)$$

The units of the calorimeter constant are cal/°C. From this definition, we may derive the calories transferred between the calorimeter and its contents as follows:

$$Q_c = K_c\, \Delta t \qquad (7.4)$$

The value of a calorimeter constant must be determined experimentally. For example, suppose 100 grams of water are placed in a calorimeter, and they come to a temperature of 30°. Now suppose that 50 grams of water at 80°C are added to the calorimeter, the contents are thoroughly mixed and they reach a final temperature of 46°C. We shall find, first, the amount of heat absorbed by the calorimeter, and then the calorimeter constant.

The heat changes in the two water samples may be calculated by Equation 7.1. These, plus the heat absorbed by the calorimeter, must equal 0, according to Equation 7.2. We can therefore say,

$$Q_{cw} + Q_{hw} + Q_c = 0 \qquad (7.5)$$

where subscripts cw, hw and c refer to cold water, hot water and calorimeter, respectively. Substituting from Equation 7.1,

$$100\text{ g} \times \frac{1.00\text{ cal}}{\text{(g) (°C)}} \times (46\text{-}30)°\text{C} + 50\text{ g} \times \frac{1.00\text{ cal}}{\text{(g) (°C)}} \times (46\text{-}80)°\text{C} + Q_c = 0$$

Solving the equation, we find $Q_c = 100$ calories. In other words, of the 1700 calories given up by the hot water [50 × 1 × (46-80)], 1600 calories went to heat the cold water [100 $\times$ 1 $\times$ (46-30)] and the remaining 100 calories were absorbed by the calorimeter. The temperature increase inside the calorimeter was from 30°C to 46°C, or $\Delta t = 16°C$. Substituting into Equation 7.3,

$$K_c = \frac{100 \text{ calories}}{16°C} = 6.25 \text{ cal/°C}$$

In our experiment, we shall first determine K_c for the calorimeter, using a hot metal of known specific heat rather than hot water. We will then find the specific heat of an unknown metal by adding a weighed mass of that metal at a known temperature to a measured quantity of water at a known lower temperature. The final temperature of the system will be measured. Then, according to Equation 7.2,

$$Q_m + Q_w + Q_c = 0$$

where Q_m is the heat lost by the metal. When we replace the Q terms by their equivalents from Equation 7.1 and 7.4, we find

$$(\text{grams} \times \text{sp. ht.} \times \Delta t)_m + (\text{grams} \times \text{sp. ht.} \times \Delta t)_w + K_c \Delta t_c = 0 \quad (7.6)$$

The only unknown in this equation is the specific heat of the metal.

One of the earliest attempts to measure atomic weight was proposed by Dulong and Petit in 1819. The essence of their proposal is that the product of atomic weight and specific heat of a solid element is a constant, its value being 6.2. From this,

$$\text{Atomic weight} = \frac{6.2}{\text{specific heat}} \quad (7.7)$$

We shall use the Dulong and Petit relationship to estimate the atomic weight of the unknown metal.

PROCEDURE

1. DETERMINATION OF CALORIMETER CONSTANT

Place about 500 ml water into a 600 ml beaker and heat it to boiling. Proceed as follows while waiting.

Obtain a Styrofoam "hot drink" cup (preferably two, one nested in the other) for your calorimeter. Weigh the cup on a centigram balance; then weigh into it about 100 g of water. Record the mass of water in the report sheet.

Weigh out 40 to 45 grams of copper shot into a preweighed large diameter test tube, using a centigram balance. Determine the actual weight of the copper shot and record it in the report sheet. Mount the test tube on a ring stand and immerse it in the boiling water in the 600 ml beaker, being sure the water level is above the level of the copper in the test tube. Allow the copper to heat for 15 minutes. Record the temperature of the boiling water.

Allow the thermometer to cool to room temperature. Then measure and record the temperature of the water in the calorimeter. Be sure the thermometer bulb is completely submerged when reading the temperature. Take the test tube from the boiling water and quickly pour the copper shot into the calorimeter, taking care not to lose any of the water by splashing. Note the time. Stir the water in the calorimeter by swirling. Avoid hitting the thermometer bulb with the copper shot. Take and record temperature readings at 15 second intervals for 3½ to 4 minutes.

Pour out the water from your calorimeter. Dry the copper shot and return them to the supply bottle.

2. SPECIFIC HEAT OF UNKNOWN METAL

Repeat the entire procedure from Part 1, this time substituting for the copper 40 to 45 grams of unknown metal supplied by the instructor.

RESULTS AND CALCULATIONS

1. DETERMINATION OF FINAL (EQUILIBRIUM) TEMPERATURE

In order to determine accurately the final temperature reached by the mixture (water + solid), we must plot the temperature readings against time and extrapolate the temperature back to the time of mixing. NOTE: "To extrapolate" means to extend observed data beyond values actually measured. This procedure is necessary because, while the temperature of the water is still rising, the system is already losing heat to the surroundings. It is therefore impossible to determine by direct measurement the final temperature (t_f) of the system.

On the graph paper provided, plot your observed temperatures vs. time. Select a temperature scale that will make the smallest division on the graph equal to 0.1°C. Extend the straight line portion of the curve back to zero time, the time when you poured the solid into the water. If the first or second points are not on the straight line portion of the graph, disregard them. They indicate simply that the water temperature was still rising while the heat was being transferred from the hot metal. The intersection of the extrapolated line with the temperature axis represents the final temperature that theoretically would have been reached if the entire heat flow from metal to water had occurred instantaneously.

Plot separate curves for Parts 1 and 2. Record the final temperatures reached.

2. CALCULATION OF THE CALORIMETER CONSTANT

From the data of the experiment, calculate K_c by direct substitution into Equation 7.6. The specific heat of copper is $0.0921 \dfrac{cal}{(g)\ (^\circ C)}$. The calorimeter constant is the only unknown. Itemize the steps as shown on the report sheet.

3. CALCULATION OF THE SPECIFIC HEAT OF THE UNKNOWN

From the data of the experiment, calculate the specific heat of the sample metal by direct substitution into Equation 7.6. The specific heat is the only unknown. Itemize the steps as shown on the report sheet.

4. ATOMIC WEIGHT OF THE UNKNOWN

From the results of the experiment, calculate the approximate atomic weight of the unknown element by direct substitution into Equation 7.7.

Name _____

Date _____ Section _____

DATA

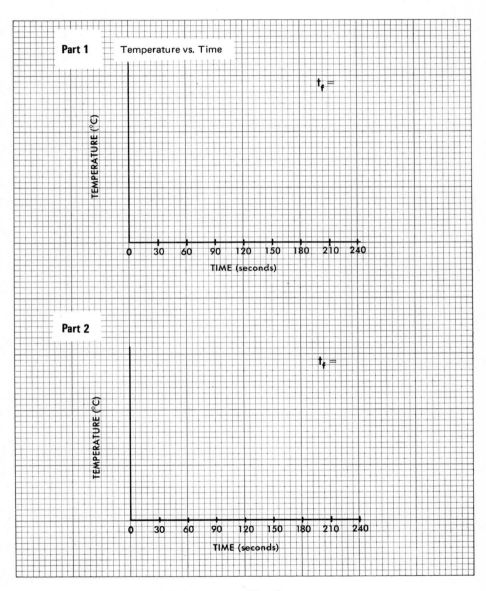

Temperature vs. Time Curves

EXPERIMENT 7
REPORT SHEET

Name _____

Date _____   Section _____

DATA

Time-Temperature Readings

Seconds	Part 1 Temperatures	Part 2 Temperatures
15	_____	_____
30	_____	_____
45	_____	_____
60	_____	_____
75	_____	_____
90	_____	_____
105	_____	_____
120	_____	_____
135	_____	_____
150	_____	_____
165	_____	_____
180	_____	_____
195	_____	_____
210	_____	_____
225	_____	_____
240	_____	_____

DATA

	Part 1: Copper	Part 2: Unknown No. ___
Mass of calorimeter	_____ g	_____ g
Mass of calorimeter and water	_____ g	_____ g
Mass of water	_____ g	_____ g
Mass of test tube	_____ g	_____ g
Mass of test tube and metal	_____ g	_____ g
Mass of metal	_____ g	_____ g
Temperature of water in calorimeter	_____ °C	_____ °C
Temperature of boiling water	_____ °C	_____ °C
Final temperature (from graph)	_____ °C	_____ °C

EXPERIMENT 7
REPORT SHEET

Name _____

Date _____ Section _____

Page 2

CALCULATIONS: (SHOW SETUPS FOR EACH CALCULATION)

Part 1

1. Heat lost by the copper shot (Q_m):

2. Heat gained by the water (Q_w):

3. Heat change of calorimeter (Q_c):

4. Calorimeter constant:

Part 2

Unknown No. _____

1. Heat gained by the water (Q_w):

2. Heat gained (or lost) by the calorimeter (Q_c):

3. Heat lost by unknown (Q_m):

4. Specific heat of the unknown:

5. Atomic weight of the unknown:

EXPERIMENT 7
ADVANCE STUDY ASSIGNMENT

Name _____

Date _____ Section _____

1. The specific heat of substance A is 0.0840 $\dfrac{cal}{(g)\,(°C)}$; that of substance B is 0.925 $\dfrac{cal}{(g)\,(°C)}$. Which substance is a better heat conductor? Explain.

2. Suppose your thermometer read 1° too high at all temperatures. Would this error make the atomic weight too large or too small? What if the thermometer read 1° too low at all temperatures? Explain.

Stoichiometry: Percent Oxygen in Potassium Chlorate

PERFORMANCE GOALS

8-1 Heat a solid to constant weight in a test tube.

8-2 Determine the percentage of one part of a compound by thermal decomposition.

CHEMICAL OVERVIEW

The thermal decomposition of potassium chlorate is described by the equation

$$2 \text{ KClO}_3 \text{ (s)} \xrightarrow{\Delta} 2 \text{ KCl (s)} + 3 \text{ O}_2 \text{ (g)} \tag{8.1}$$

In this experiment, you will confirm the stoichiometry of this reaction and determine the percentage of oxygen in potassium chlorate. *Stoichiometry* refers to calculations involving mass and mole relationships in chemical reactions.

While the potassium chlorate will decompose simply by heating, the reaction is intolerably slow. A catalyst, manganese dioxide, MnO_2, is therefore added to speed the reaction. Though it contains oxygen, the catalyst experiences no permanent change during the reaction and does not contribute measurably to the oxygen generated. As with all catalysts, the quantity present at the end of the reaction is the same as at the beginning.

The procedure here is to weigh a quantity of potassium chlorate, heat it to drive off the oxygen and then weigh the residue, which is assumed to be potassium chloride. The loss in weight represents the oxygen content of the original potassium chlorate.

The detailed procedure first involves weighing the test tube, test tube holder and dry catalyst, then adding potassium chlorate and weighing again to determine the weight of the KClO_3 by difference. The oxygen is then driven off, the test tube and its contents weighed again and the weight of the KCl determined by difference.

In a thermal decomposition such as this, the product must be "heated to constant weight" before you can be sure the decomposition is complete. After the first heating, cooling and weighing of the decomposed product, the test tube is heated a second time and is cooled and weighed again. If the two weighings are the same, within the experimental limits of the equipment used, it may be assumed that all of the oxygen was removed in the first heating. If weight is lost in the second heating, it means that some oxygen remained after the first heating and was driven off in the second. It is possible that some oxygen may still be present after the second heating. The procedure is therefore repeated again, as many times as necessary, until there is no loss is weight between two consecutive weighings. Hence the expression *heating to constant weight.*

PROCEDURE

Select a medium-size test tube and add 0.5 to 0.8 grams of MnO_2. Heat the test tube over a Bunsen burner for about 3 to 4 minutes to drive off any moisture that may be present in the catalyst and the test tube. Allow the test tube and its contents to cool to room temperature. Hang the test tube, test tube holder and catalyst from the hook above the pan of a centigram balance, weigh it, and record the weight. This dead weight will be present in all subsequent weighings. Add from 1.0 to 1.5 grams of $KClO_3$, mix the contents thoroughly until they have a uniform gray appearance (be careful not to lose any of the contents of the test tube) and weigh again.

CAUTION: THE NEXT STEP IS POTENTIALLY HAZARDOUS. THE FORMA-
TION OF A GAS AT THE BOTTOM OF A TEST TUBE MAY RESULT IN A
SUDDEN EXPANSION, BLOWING HOT CHEMICALS OUT OF THE TEST
TUBE. THIS MISHAP WILL NOT OCCUR IF THE TEST TUBE IS HANDLED
PROPERLY DURING HEATING. FOLLOW THE NEXT INSTRUCTIONS
PRECISELY. BE ABSOLUTELY SURE YOUR TEST TUBE IS NOT POINTING
AT ANYBODY AS IT IS HEATED.

Tip your test tube until it is almost horizontal and tap it carefully until the contents are distributed over the lower half of the length of the tube, as shown in Figure 8-1. Holding it at about this angle, move the test

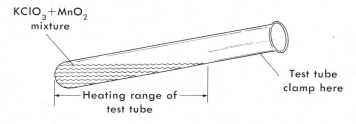

Figure 8-1.

tube back and forth in the flame of a Bunsen burner, distributing the heat over the entire length of the mixture. Heat gently at first, increasing the intensity after the mixture seems to "boil," as it sometimes appears to do when bubbles of oxygen are being released. Continue heating for about five minutes, cool and weigh. Repeat the process as necessary to obtain constant weight. *Record each weighing* in your report sheet.

CALCULATIONS

Using only recorded weighings, determine the following by difference: (a) the starting quantity of potassium chlorate; (b) the final quantity of potassium chloride; (c) the total mass of oxygen lost in heating. (Careful on the last one; use only *recorded weighings.*) Show each subtraction.

Based on the quantities found above, calculate the experimental percentage of oxygen in potassium chlorate. Compare this value with the theoretical percent of oxygen calculated from the formula.

EXPERIMENT 8
REPORT SHEET

Name _____

Date _____ Section _____

DATA

Weight of test tube, test tube holder and dried MnO_2 _____ g

Weight of test tube holder, MnO_2 and $KClO_3$ _____ g

Weight after first heating and cooling _____ g

Weight after second heating and cooling _____ g

Weight after third heating and cooling (if necessary) _____ g

Weight after fourth heating and cooling (if necessary) _____ g

Weight after fifth heating and cooling (if necessary) _____ g

RESULTS

Weight of $KClO_3$ used (show calculations):

_____ g

Weight of KCl after heating to constant weight (show calculations):

_____ g

Weight of oxygen released (show calculations):

_____ g

Percent oxygen in potassium chlorate (show both calculations):

Experimental:

_____ %

Theoretical:

_____ %

EXPERIMENT 8

ADVANCE STUDY ASSIGNMENT

Name _____

Date _____ Section _____

1. What is a catalyst? How does a catalyst affect a chemical reaction?

2. What is the meaning of "heating to constant weight"? Why is it done?

3. Calculate the theoretical percent oxygen in $KClO_3$.

Names and Formulas of Compounds: A Study Assignment

PERFORMANCE GOALS

9-1 Given the name (or formula) of an ionic or binary covalent compound, write its formula (or name).

INTRODUCTION

In this study assignment only a brief summary of rules for writing formulas and naming compounds is presented. You should consult your textbook for more complete and detailed information. The purpose of this exercise is to provide a practice session in writing formulas and names, with help immediately available to clear up points that you may not understand. Hopefully, you will *master* formula writing techniques during this laboratory period.

CHEMICAL OVERVIEW

Oxidation Numbers

The oxidation number of an element is a positive or negative number assigned to it according to a system based on how that element enters into compounds. In an ionic compound, the oxidation number of an atom or group of atoms existing as an ion (polyatomic ion) equals the charge on that ion. In covalent compounds, oxidation numbers are arbitrarily assigned to each atom based on the number of electrons shared. Rules governing oxidation numbers are summarized below:

1. The oxidation number of a free element is zero.
2. The oxidation number of an element in a monatomic ion is equal to the charge on the ion.
3. The oxidation number of oxygen is -2 except in peroxides.
4. The oxidation number of hydrogen is $+1$ except in hydrides.

5. The algebraic sum of oxidation numbers for all atoms in a compound is zero.
6. The algebraic sum of oxidation numbers for all atoms in a polyatomic ion (e.g., NH_4^+, SO_4^{2-}) is equal to the charge on the ion.

We can predict the oxidation number of some of the elements from their electron structure. Positive ions (cations) derived from the IA, IIA and IIIA groups of the periodic table have only one possible oxidation number. The other A group elements may have more than one oxidation number when covalently combined with other atoms. Table 9-1 lists the predictable oxidation numbers for the A group elements.

TABLE 9-1. OXIDATION NUMBERS

Group	IA	IIA	IIIA	IVA	VA	VIA	VIIA
Monatomic Ions	+1	+2	+3			-2	-1
Covalently Bonded Nonmetals				-4 to +4	-3 to +5	-2 to +6	-1 to +7

Predictions for the transition elements (B groups and Group VIII in the periodic table) are more complicated and will not be made in this exercise.

Writing Formulas of Ionic Compounds

Ionic compounds contain positively and negatively charged ions in proportions such that the sum of the charges equals zero.

TABLE 9-2. IONIC FORMULAS

Ions		Charges	Compound	Sum of Charges
K^+	Cl^-	+1, -1	KCl	$+1 + (-1) = 0$
Ba^{2+}	S^{2-}	+2, -2	BaS	$+2 + (-2) = 0$
Ca^{2+}	Br^-	+2, -1	$CaBr_2$	$+2 + 2(-1) = 0$
Fe^{3+}	O^{2-}	+3, -2	Fe_2O_3	$2(+3) + 3(-2) = 0$

Rules for writing the formulas of ionic compounds are as follows:

1. Write the positive ion first, the negative ion second.
2. Write the numeral indicating the number of ions, if other than 1, as a subscript following the symbol of the ion.
3. If a polyatomic ion appears more than once in a formula, enclose it in parentheses.
4. Subscripts in almost all compounds are in the lowest possible whole number ratio of the atoms.

When writing the formula of an ionic compound, always remember to check that the sum of the ionic charges is zero; or, to put it another way, be sure that the numerical sum of the positive charges equals the numerical sum of the negative charges.

Naming Ionic Compounds

In naming ionic compounds, you name the positive ion first, followed by the name of the negative ion — just the way the formula is written. If the positive ion may exist in more than one oxidation state, either the classical *-ous, -ic* system or the officially accepted Stock System (accepted by the International Union of Pure and Applied Chemistry, sometimes called the IUPAC system) is used. The classical nomenclature is limited to those metals in which the ions have only two possible oxidation states, such as iron and copper. The ion having the lower oxidation number is given the *-ous* ending, and the higher one the *-ic* ending. Thus, Fe^{2+} is ferrous, and Fe^{3+} is ferric; Cu^+ is cuprous, and Cu^{2+} is cupric. Mercury is unique. In addition to a typical monatomic ion with a +2 oxidation state, it forms a diatomic ion with a +1 oxidation state. The mercurous ion is $Hg_2{}^{2+}$, and the mercuric ion is Hg^{2+}.

The Stock System, which is now more widely used and growing in popularity, employs Roman numerals in parentheses immediately following the symbol of the element to indicate its oxidation number. Both systems are summarized in the examples that follow:

TABLE 9-3. NAMES OF IONIC COMPOUNDS

Formula	Classical Name	IUPAC Name
$FeCl_2$	Ferrous chloride	Iron (II) chloride
$FeCl_3$	Ferric chloride	Iron (III) chloride
CuCl	Cuprous chloride	Copper (I) chloride
$CuCl_2$	Cupric chloride	Copper (II) chloride

Oxyacids and Their Salts

Compounds containing hydrogen, oxygen and one other element are called **oxyacids**. The third element and the number of oxygen atoms in the molecule give the acid its name. For example, $HClO_3$ is *chlor*ic acid, from *chlor*ine; $HBrO_3$ is *brom*ic acid, from *brom*ine; and HNO_3 is *nitr*ic acid, from *nitr*ogen. The number of oxygen atoms is conveyed by a prefix or a suffix, or both. A similar system establishes the names of the anions derived from oxyacids. Table 9-4 summarizes the naming of the most common oxyacids and their salts.

TABLE 9-4. NAMES OF OXYACIDS AND THEIR SALTS

Acid Name	Cl	S	N	P	C	Anion
per ... ic	$HClO_4$					per ... ate
... ic	$HClO_3$	H_2SO_4	HNO_3	H_3PO_4	H_2CO_3	... ate
... ous	$HClO_2$	H_2SO_3	HNO_2	H_3PO_3		... ite
hypo ... ous	$HClO$			H_3PO_2		hypo ... ite

When the hydrogen of an acid is replaced by a metal ion (or ammonium ion), a salt is formed. When prefixes such as *hypo-* or *per-* are used for the acid, they are also used for its salt. The name of the anion in the salt depends on the acid from which it is derived. As shown in Table 9-4, an anion has an -*ate* ending if it comes from an acid ending in -*ic*; and the anion ends in -*ite* if the parent acid ends in -*ous*. Table 9-5 lists several examples:

TABLE 9-5. NAMES OF SALTS

Salt	Parent Acid	Name of Salt
Na_2SO_4	H_2SO_4	Sodium sulfate
$KClO_2$	$HClO_2$	Potassium chlorite
$NaBrO$	$HBrO$	Sodium hypobromite
Li_3PO_3	H_3PO_3	Lithium phosphite
$Ca(ClO_4)_2$	$HClO_4$	Calcium perchlorate

Acid Salts. Acid salts are compounds obtained by partial substitution of metals for hydrogen in an oxyacid containing more than one hydrogen. Examples are $NaHCO_3$, sodium hydrogen carbonate; $KHSO_3$, potassium hydrogen sulfite; Na_2HPO_4, sodium monohydrogen phosphate; and KH_2PO_4, potassium dihydrogen phosphate.

Covalent Binary Compounds

Compounds containing two nonmetals held together by a covalent bond (sharing of electrons) are called **covalent binary compounds.** In such compounds, the element having the lower electronegativity is named first and is written first in the formula.* The following list arranges the most important nonmetals in order of increasing electronegativity:

$$\text{Si, P, H, C, S, I, Br, N, Cl, O, F}$$

Prefixes are used to indicate the number of atoms of each element in the compound. The prefix is commonly omitted if there is only one atom in the molecule. The most common prefixes and their numerical equivalents are shown below:

TABLE 9-6. PREFIXES USED IN NAMING COVALENT BINARY COMPOUNDS

mono-	= 1	penta-	= 5
di-	= 2	hexa-	= 6
tri-	= 3	hepta-	= 7
tetra-	= 4	octa-	= 8

Examples of covalent binary compounds and their names are listed in Table 9-7:

TABLE 9-7. COVALENT BINARY COMPOUNDS

Formula	Name
PCl_3	Phosphorus trichloride
NO_2	Nitrogen dioxide
CBr_4	Carbon tetrabromide
CS_2	Carbon disulfide
BF_3	Boron trifluoride
N_2O_4	Dinitrogen tetroxide

---

*Hydrogen is sometimes an exception to this rule, as in methane, CH_4, and ammonia, NH_3.

EXPERIMENT 9
REPORT SHEET

Name _____

Date _____ Section _____

A. Ionic Compounds

a. Write formulas for the following compounds:

    1. Sodium nitrate   _____

    2. Calcium fluoride   _____

    3. Sodium carbonate   _____

    4. Potassium iodide   _____

    5. Magnesium chloride   _____

    6. Ammonium sulfate   _____

    7. Barium hydroxide   _____

    8. Silver bromide   _____

    9. Potassium nitrite   _____

  10. Calcium chlorate   _____

  11. Aluminum sulfite   _____

  12. Magnesium oxide   _____

  13. Sodium hypochlorite   _____

  14. Ammonium carbonate   _____

  15. Barium phosphite   _____

b. Name the following compounds:

1. K_2SO_4 _____

2. Na_3PO_4 _____

3. NH_4Br _____

4. $Ca(OH)_2$ _____

5. $BaCl_2$ _____

6. $NaClO_2$ _____

7. K_2O _____

8. $AgBrO_3$ _____

9. CaI_2 _____

10. KIO_3 _____

11. $Al_2(SO_3)_3$ _____

12. Ag_2SO_4 _____

13. $NaClO$ _____

14. K_3PO_3 _____

15. $(NH_4)_2CO_3$ _____

EXPERIMENT 9
REPORT SHEET

Name _____

Date _____ Section _____

Page 2

B. Ionic Compounds Containing Cations of Varying Oxidation Numbers

a. Write formulas for the following compounds:

1. Iron (III) sulfide _____

2. Lead (II) nitrate _____

3. Copper (I) iodate _____

4. Mercury (II) sulfate _____

5. Iron (III) hydroxide _____

6. Copper (I) nitrite _____

7. Iron (II) sulfite _____

8. Mercury (II) chlorate _____

9. Copper (II) carbonate _____

10. Copper (I) bromide _____

b. Name the following compounds:

1. $FeCl_3$ _____

2. $CuSO_4$ _____

3. Fe_2O_3 _____

4. PbI_2 _____

5. $CuBrO_3$ _____

6. $FePO_4$ _____

7. $HgCO_3$ _____

8. $Fe(OH)_2$ _____

9. Cu_2O _____

10. $PbSO_3$ _____

EXPERIMENT 9
REPORT SHEET

Name _____

Date _____ Section _____

Page 3

C. Oxyacids

a. Write formulas for the following compounds:

  1. Hypochlorous acid _____

  2. Sulfuric acid _____

  3. Nitrous acid _____

  4. Perbromic acid _____

  5. Iodous acid _____

  6. Phosphoric acid _____

  7. Sulfurous acid _____

  8. Chloric acid _____

b. Name the following compounds:

1. H_3PO_3 _____

2. HNO_3 _____

3. H_2SO_3 _____

4. $HBrO_2$ _____

5. HIO _____

6. $HClO_4$ _____

7. H_3PO_4 _____

8. HIO_3 _____

EXPERIMENT 9
REPORT SHEET

Name _____

Date _____ Section _____

Page 4

D. Acid Salts

a. Write formulas for the following compounds:

1. Sodium hydrogen carbonate _____

2. Sodium monohydrogen phosphate _____

3. Potassium hydrogen sulfite _____

4. Sodium dihydrogen phosphite _____

5. Calcium hydrogen sulfate _____

b. Name the following compounds:

1. $NaHSO_4$ _____

2. $Ba(HSO_3)_2$ _____

3. Na_2HPO_3 _____

4. KH_2PO_4 _____

5. $NaHCO_3$ _____

E. Covalent Binary Compounds

a. Write formulas for the following compounds:

1. Diphosphorus pentoxide _____

2. Dinitrogen tetroxide _____

3. Iodine chloride _____

4. Carbon monoxide _____

5. Sulfur trioxide _____

b. Name the following compounds:

1. CO_2 _____

2. NO _____

3. P_2O_3 _____

4. SO_2 _____

5. CCl_4 _____

EXPERIMENT 9
REPORT SHEET

Name _____

Date _____ Section _____

Page 5

F. Compounds Containing Less Common Ions

(Use Table of anions and cations in Appendix and refer to the Periodic Table.)

Write formulas for the following compounds:

1. Strontium chloride (strontium, atomic no. 38)   _____

2. Cesium bromide (cesium, atomic no. 55)   _____

3. Potassium chromate   _____

4. Sodium cyanide   _____

5. Aluminum borate   _____

6. Silicon dioxide (silicon, atomic no. 14)   _____

7. Calcium hydride   _____

8. Manganese dioxide   _____

9. Selenium dichloride (selenium, atomic no. 34)   _____

10. Sodium thiocyanate   _____

EXPERIMENT 9
ADVANCE STUDY ASSIGNMENT

Name _____

Date _____   Section _____

1. Describe the use of the Stock System for the naming of cations having more than one oxidation state. Give two examples.

2. Give the names and formulas of two oxyacids of chlorine.

3. What are the endings given to salts derived from (a) -ous acids and (b) -ic acids? Give two examples.

4. What is the formula of the salt derived from the cation X^{3+} and anion Y^{2-}?

Qualitative Analysis of Some Common Ions

PERFORMANCE GOALS

10-1 Conduct tests to confirm the presence of known ions in a solution.
10-2 Analyze an unknown solution for certain ions.

CHEMICAL OVERVIEW

When we analyze an unknown solution, two questions come to mind: (a) what ions are present in the solution, and (b) what is their concentration? The first question can be answered by performing a **qualitative analysis**, and the second by a **quantitative analysis**. These two broad categories are known collectively as **analytical chemistry**. In this experiment, we shall perform a qualitative analysis.

The general approach to finding out what ions are in a solution is to test for the presence of each possible component by adding a reagent that will cause that component, if present, to react in a certain way. This method involves a series of tests, one for each component, carried out on separate samples of solution. Difficulty sometimes arises, particularly in complex mixtures, because one species may interfere with the analytical test for another. Although interferences are common, many ions in mixtures can usually be identified by simple tests.

In this experiment, we shall analyze an unknown mixture that may contain one or more of the following ions in solution:

$$CO_3{}^{2-} \quad Cl^- \quad SCN^- \quad SO_4{}^{2-} \quad CrO_4{}^{2-}$$

$$C_2H_3O_2{}^- \quad NH_4{}^+ \quad Cu^{2+} \quad Ni^{2+}$$

PROCEDURE

A boiling water bath is required for some of the tests you are to perform. Pour about 100 milliliters of tap water into a 150 milliliter beaker and heat it to boiling. Maintain it at that temperature throughout the experiment, replenishing the water from time to time as it becomes necessary.

1. TEST FOR THE CARBONATE ION, $CO_3{}^{2-}$

Cautiously add 1 milliliter of 6 M HCl to 1 milliliter of 1 M Na_2CO_3 in a medium-size test tube. Bubbles of carbon dioxide usually appear immediately in the presence of the carbonate ion. If the bubbles are not readily apparent, warm the solution in the hot water bath and stir.

2. TEST FOR THE ACETATE ION, $C_2H_3O_2{}^-$

Cautiously add 1 milliliter of 6 M HCl to 1 milliliter of 0.5 M $NaC_2H_3O_2$ and stir. If acetate ion is present, acetic acid, with the characteristic odor of vinegar, is formed. Heating for 30 seconds in the hot water bath intensifies the odor.

3. TEST FOR THE SULFATE ION, $SO_4{}^{2-}$

Cautiously add 1 milliliter of 6 M HCl to 1 milliliter of 0.5 M Na_2SO_4. Then add 3 to 4 drops of 1 M $BaCl_2$. A white, powdery precipitate of $BaSO_4$ indicates the presence of $SO_4{}^{2-}$ ions in the sample.

4. TEST FOR CHROMATE ION, $CrO_4{}^{2-}$

Solutions containing chromate ion are yellow when neutral or basic and orange when acidic. Add 2 milliliters of 6 M HNO_3 to 1 milliliter of 0.5 M K_2CrO_4. If a color change to pale blue occurs shortly after the addition of the acid, the presence of chromate ion is confirmed. If no color change is observed, cool the test tube under tap water and add 1 milliliter of 3% hydrogen peroxide, H_2O_2. If chromate ion is present, you will observe a rapidly fading blue color caused by the formation of unstable chromium peroxide, CrO_5.

5. TEST FOR THIOCYANATE ION, SCN^-

Add 1 milliliter of 6 M $HC_2H_3O_2$ to 1 milliliter of 0.5 M KSCN and stir with a glass rod. Add 1 or 2 drops of 0.1 M $Fe(NO_3)_3$. A deep red color formation is proof of the presence of SCN^- ions.

6. TEST FOR CHLORIDE ION, Cl^-

Add 1 milliliter of 6 M HNO_3 to 1 milliliter of 0.5 M NaCl. Add 2 or 3 drops of 0.1 M $AgNO_3$. A white precipitate of AgCl confirms the presence of chloride ion.

If thiocyanate ion is present, it will interfere with this test, since it also forms a white precipitate with $AgNO_3$. If the sample contains SCN^- ion, put 1 milliliter of the solution in a small 30 or 50 milliliter beaker and add 1 milliliter of 6 M HNO_3. Boil the solution gently until the volume is reduced to half; this procedure will oxidize the thiocyanate and remove the interference. Then perform the chloride ion test as previously explained.

7. TEST FOR THE AMMONIUM ION, NH_4^+

Add 1 milliliter of 6 M NaOH to 1 milliliter of 0.5 M NH_4Cl. The ammonium ion reacts with the hydroxide to form NH_3, which can be detected by its characteristic odor. For a more sensitive test, place 1 milliliter of the ammonium chloride solution into a small beaker. Moisten a piece of red litmus paper and put it on the bottom of a watch glass. Cover the beaker with the watch glass and gently heat the liquid to the boiling point, but do not boil it. Take care that no liquid comes into contact with the litmus paper. If NH_4^+ ion is present, the litmus paper gradually turns blue on exposure to the evolving NH_3 vapors.

8. TEST FOR THE NICKEL (II) ION, Ni^{2+}

Add 2 or 3 drops of 1 M NH_3 to 1 milliliter of 0.5 M $NiSO_4$. Then add 2 milliliters of 1% alcoholic dimethylglyoxime and stir with a glass rod. A red, gelatinous precipitate forms if Ni^{2+} is present. The precipitate is soluble in dilute HCl. To test for Ni^{2+} in an unknown solution, make the solution slightly alkaline with dilute ammonia. Then proceed with the addition of dimethylglyoxime, as previously described.

9. TEST FOR THE COPPER (II) ION, Cu^{2+}

Add concentrated ammonia drop by drop to 1 milliliter of 0.5 M $CuSO_4$. The development of a dark blue color, due to a complex ion of copper, is proof of the presence of copper(II) ion.

10. ANALYSIS OF AN UNKNOWN

When you have completed all of the tests, obtain an unknown from the instructor and analyze it by dividing the unknown into 1 milliliter portions and then applying the tests to separate portions. The unknown will contain three to five of the ions on the list, so your test for a given ion may be affected by the presence of others. When a test does not go quite according to the rules, try to deduce why the sample may have behaved as it did. When you think you have analyzed your unknown properly, you may, if you wish, make a "known" that has the composition you found and test it to see if it has the properties of the unknown.

Note: In case the test for Ni^{2+} is positive, an additional test for Cu^{2+} should be performed by adding 2 milliliters of thioacetamide to the *acidic* unknown solution. Precipitate formation indicates the presence of Cu^{2+}.

**EXPERIMENT 10
REPORT SHEET**

Name _____

Date _____   Section _____

Unknown Number: _____

Ion Tested	Observation and Comment	
	Known	Unknown
CO_3^{2-}		
$C_2H_3O_2^-$		
SO_4^{2-}		
CrO_4^{2-}		
SCN^-		
Cl^-		
NH_4^+		

Ion Tested	Observation and Comment	
	Known	Unknown
Ni^{2+}		
Cu^{2+}		

Ions present in unknown number _____ : _____

EXPERIMENT 10
ADVANCE STUDY ASSIGNMENT

Name _____

Date _____ Section _____

1. An unknown that might contain any of the nine ions studied in this experiment has the following properties:
   a. The unknown is colorless and odorless.
   b. On adding 6 M HCl and warming the unknown, you detect the odor of vinegar.
   c. When 0.1 M $BaCl_2$ is added to the acidified unknown, a clear solution results.
   d. When 6 M NaOH is added to the unknown, the resulting vapor turns moist red litmus paper blue.
   On the basis of the preceding information, classify each of the following ions as present (P), absent (A) or undetermined (U) by the tests described:
   $CO_3{}^{2-}$ _____ ; Cl^- _____ ; $SO_4{}^{2-}$ _____ ; $C_2H_3O_2{}^-$ _____ ; $CrO_4{}^{2-}$ _____ ; $NH_4{}^+$ _____ ; Cu^{2+} _____ ; Ni^{2+} _____ .

Optional Assignment

Write net ionic equations for the reactions in this experiment in which the following ions are detected:

a. $CO_3{}^{2-}$: _____

b. $C_2H_3O_2{}^-$: _____

c. $SO_4{}^{2-}$: _____

d. Cl^-: _____

e. $NH_4{}^+$: _____

Boyle's Law

PERFORMANCE GOALS

11-1 Given the atmospheric pressure, use an open end manometer to measure the pressure exerted by a confined gas.

11-2 State Boyle's Law and its mathematical significance.

CHEMICAL OVERVIEW

Boyle's Law states the relationship between the volume and pressure of a fixed quantity of confined gas at constant temperature. In performing this experiment, you will measure pressure and volume in a rather simple apparatus consisting of two glass tubes connected by a piece of rubber tubing, as shown in Figure 11-1. (More elaborate commercial apparatus operates under the same principle.) One glass tube is open to the atmosphere, while the other is closed by a screw clamp applied to a rubber hose just above the end of the glass. The system contains mercury. The surface of the mercury in the open tube is subject to atmospheric pressure, whereas the surface of the mercury in the closed tube is subject to the pressure applied by the gas trapped between the surface and the point where the tube is clamped shut. The volume occupied by this trapped air is obviously the volume of the tube between these points. These are the pressures and volumes to be measured in this experiment.

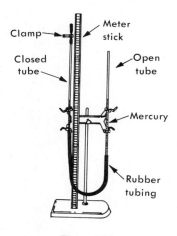

Figure 11-1.

In measuring the volume of the confined gas, we make the reasonable assumption that the cross-sectional area of the glass tube is constant throughout its length. The volume is then equal to the product of this area and the length of the trapped air column. Because the area is constant, it follows that the volume is proportional to the length of the air space and that the length is effectively a measure of the volume. Volume will be expressed in terms of this length, measured in millimeters.

The apparatus is essentially an open end manometer. The method of measuring pressure with such manometers is based on the equality of pressures in the two legs of the manometer *at the level of the lower liquid surface.* These pressures are the pressures exerted by the gas, or the gas and the liquid, above this level. Two possible conditions exist: (a) where the mercury level is higher in the open tube (Figure 11-2A) and (b) where the mercury level is lower in the open tube (Figure 11-2B). In the first case, Figure 11-2A, the lower liquid surface is in the closed tube. The only pressure applied at that point is the pressure of the gas, P_g. *At the same level* in the open tube, the pressure is the result of everything above that point, or $P_a + P_{Hg}$, the sum of the pressure of the atmosphere plus the pressure exerted by the mercury column. Equating these pressures at the lower liquid level, we have

$$P_g = P_a + P_{Hg} \tag{11.1}$$

P_a may be determined by reading a barometer. P_{Hg} is simply the height of the mercury column. Both pressures should be measured in millimeters of mercury.

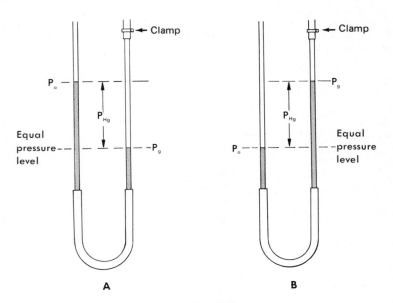

Figure 11-2.

In the second case, equating pressures at the lower liquid level gives

$$P_g + P_{Hg} = P_a \qquad (11.2)$$

Solving for P_g,

$$P_g = P_a - P_{Hg} \qquad (11.3)$$

When we compare Equations 11.1 and 11.3, it is evident that the pressure of the gas is equal to the pressure of the atmosphere, plus or minus the height of the mercury column. If the gas pressure is greater than atmospheric pressure (Figure 11-2A), the length of the mercury column is added to atmospheric pressure; if the gas pressure is less than atmospheric, the mercury column length is subtracted.

PROCEDURE

Record the atmospheric pressure on the report sheet. If it has not already been done, adjust the clamped hose so that it is at the level of zero at the top of the meter stick. With this arrangement, the volume (length) may be read directly. The meter stick should be mounted next to the closed tube and should remain in fixed position throughout the experiment.

Remove the open tube from its clamp and hold it next to the meter stick, but on the opposite side from the closed tube. Move it up and down until the two mercury levels coincide. Record this measurement in millimeters in both the "Open" and "Closed" columns on the report sheet.

Raise the open tube as high as the apparatus will permit without pinching the rubber hose or causing the mercury level to rise above the top of the meter stick. Hold the tube next to the meter stick and record the mercury levels in both open and closed tubes. (When you raised the open tube, its mercury surface went up. Which way did the mercury surface move in the closed tube?)

Now lower the open tube as far as the apparatus permits. Again record the open and closed tube readings.

CAUTION: THE LIMIT OF THE EQUIPMENT MAY NOT BE SET BY THE RUBBER TUBE OR METER STICK THIS TIME, BUT RATHER BY THE LEVEL AT WHICH THE MERCURY BEGINS TO SPILL OUT OF THE TUBE. STAY COMFORTABLY ABOVE THIS LEVEL. IF YOU SHOULD HAPPEN TO SPILL ANY MERCURY, AT THIS OR ANY OTHER POINT, NOTIFY THE INSTRUCTOR IMMEDIATELY. THOUGH FUN TO PLAY WITH, MERCURY MUST BE RECOGNIZED AS A DEADLY POISON, AND SPILLED MERCURY MUST NOT BE PERMITTED TO REMAIN IN LITTLE UNSEEN AND UN-REPORTED POOLS IN THE LABORATORY. FURTHERMORE, IT IS PROBABLY MORE DEADLY AT HOME, WHERE YOUR VENTILATION IS NOT AS GOOD AS IN THE LABORATORY. NO SOUVENIRS, PLEASE!

Select two more positions for the open tube, one between the first and second positions, and the other between the first and third, recording open and closed tube readings for each. Be sure these positions are not right next to one of your earlier readings. You will now have readings for a total of five positions.

CALCULATIONS

In the data and result table, the column headed DIFFERENCE refers to P_{Hg} in Equations 11.1 and 11.3. P_{Hg} is simply the height of the mercury column in millimeters, determined from your open and closed tube readings. From this difference and from your determination of atmospheric pressure, the pressure exerted by the gas may be calculated.

Volume is expressed in terms of the height of the gas column, as explained in the overview.

When pressure and volume have both been recorded, multiply them and enter the product in the P X V column.

EXPERIMENT 11
REPORT SHEET

Name _____

Date _____   Section _____

DATA AND RESULTS

Closed Tube (mm)	Open Tube (mm)	Difference (mm)	Pressure (mm Hg)	Volume (mm)	P × V

Atmospheric pressure = _____ mm Hg (from barometer reading).

1. Summarize the results of your experiment in one sentence, being sure to include any limitations that may apply.

2. Boyle's Law is frequently stated in the form of an equation:

$$P \times V = \text{constant}$$

State your results in this form.

EXPERIMENT 11
ADVANCE STUDY ASSIGNMENT

Name _____

Date _____ Section _____

1. Identify the gas measurements that are related by Boyle's Law.

2. If atmospheric pressure is 751 mm Hg, calculate the pressure of the confined gas in each case that follows.

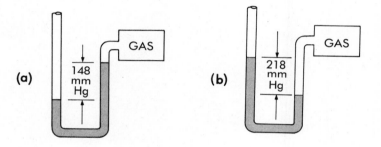

Charles' Law

PERFORMANCE GOALS

12-1 Measure the volume of a fixed quantity of a gas at different temperatures and constant pressure.

12-2 Plot a graph of the volume of gas versus the temperature at constant pressure.

12-3 Find the value of "absolute zero" from your experimental observations.

CHEMICAL OVERVIEW

Matter can exist in three states — solid, liquid or gas. Of the three states, the gaseous state is the least compact and most energetic. According to the Kinetic Molecular Theory, gas molecules move with high velocities, colliding frequently with each other and with the walls of the vessel containing them. The temperature of a gas is a measure of the average translational kinetic energy (K.E.) of the gas molecules. Kinetic energy is expressed by the equation

$$K.E. = \frac{1}{2}mv^2 \tag{12.1}$$

where m = mass of the particle and v = velocity.

If these statements and the Kinetic Molecular Theory are correct, we should be able to make some predictions about the temperature of a gas and the volume it occupies at constant pressure. Pressure is caused by collisions between the gas molecules and the walls of the container. The magnitude of pressure depends upon (1) the frequency of those collisions and (2) their "strength" or "force." If we decrease the temperature or kinetic energy of a gas, we must reduce either the mass of the individual particles or their velocity. Obviously we cannot change the particle mass, so the effect of cooling must be to reduce molecular velocity. This change, in turn, decreases both the frequency and strength of molecular collisions with the walls. We therefore predict that a decrease in temperature results in reduced pressure.

Our concern in this experiment, however, is the relationship between temperature and *volume* at *constant pressure*. If at constant volume a

111

lower pressure will result from a reduction in temperature, what change in volume will have to occur at the reduced temperature to raise the pressure back to its original value? A reduction in volume will do nothing to relieve the smaller force as the particles collide with the walls, but it will reduce the distance to be travelled between collisions and will therefore increase the collision frequency. From this consideration, it is reasonable to expect that a reduction in temperature, accompanied by a reduction in the volume of a confined gas, could maintain pressure at a constant value. We may therefore predict some sort of direct relationship between temperature and volume at constant pressure.

This line of reasoning leads to one other interesting bit of speculation. Is it possible to lower temperature to such a point that the pressure will drop to zero? We might imagine that, physically, zero pressure is the condition at which molecular motion stops altogether — particle velocity, v, becomes zero. If $v = 0$, then $\frac{1}{2}mv^2 = 0$. If temperature is a measure of average kinetic energy, and kinetic energy equals zero, then temperature must also be zero! And this is an *absolute zero,* not a zero based on some arbitrarily selected phenomenon, such as the freezing temperature of water.

In this experiment, we shall trap a fixed quantity of air between the sealed end of a piece of fine-bore glass tubing and a "plug" of mercury that is free to move up and down the tubing. If the tube is held vertically, the plug will adjust itself to a position in which the pressure of the gas exactly balances the pressure of the atmosphere plus the small additional pressure exerted by the mercury plug. In this way, the pressure of the confined gas remains constant throughout the experiment.

The volume of the confined gas, V, or the volume of the air space between the mercury plug and the sealed end of the tube, is found by multiplying the cross-sectional area of the tube, A, by the height, h, of the gas column.

$$V = A \times h \tag{12.2}$$

Assuming a constant cross-sectional area, volume is proportional to height:

$$V \propto h \tag{12.3}$$

The height of the air column may therefore be used as a measure of volume in this experiment. Measuring this height at different temperatures, we shall determine experimentally the relationship between temperature and volume of a confined gas at constant pressure, a relationship known as Charles' Law. We shall also use the data of the experiment to estimate the Celsius equivalent of "absolute zero" in temperature.

PROCEDURE

Obtain a 15-cm-long, narrow-bore glass tube (2 mm I.D.) sealed at one end and, heating the *sealed end first,* pass it through a medium-hot burner flame. The heating process will warm the air in the tube and remove any water vapor that may be present.

Hold the warm tubing with a towel and take it to the hood. Immerse the open end of the tube in the pool of mercury contained in a small beaker.

CAUTION: MERCURY IS A HIGHLY POISONOUS SUBSTANCE WHICH CAN BE ABSORBED THROUGH THE SKIN. ALSO, MERCURY TENDS TO VAPORIZE AT ELEVATED TEMPERATURES, RELEASING POISONOUS FUMES. **DO NOT SPILL OR HANDLE MERCURY!** USE IT ONLY IN THE FUME HOOD AND REPORT ANY ACCIDENTAL SPILLAGE *IMMEDIATELY* TO THE INSTRUCTOR.

As the air cools, the mercury is drawn into the tube. When the height of the mercury "plug" reaches 3/8 to 1/2 inch, withdraw the tube *gently* from the mercury reservoir and allow it to cool to room temperature. The mercury plug should be positioned about halfway down the tube. Handle the tube with care; do not shake or jar it, which could result in separation of the mercury column. Parts of a separated mercury column may be united, or the position of the entire plug changed, by inserting a fine copper wire into the mercury.

Obtain a thermometer from the instructor and fasten the glass tube to it (sealed end pointing down) with a rubber band. The bulb of the thermometer should be about at the midpoint of the trapped air column (Figure 12-1).

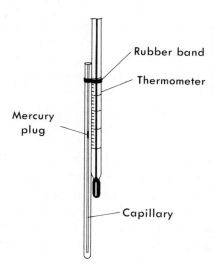

Rubber band

Thermometer

Mercury plug

Capillary

Figure 12-1.

Fill a 100 ml graduated cylinder with warm tap water (40 to 50°C) and immerse the tube and thermometer in it until the whole length of trapped air column is below the level of the water. The mercury plug will move up in the tube. When it stops, wait two minutes to be sure the gas has reached the temperature of the water. Record the temperature. *Quickly* withdraw the tube, note the exact position of the bottom of the plug and measure the height of the air column from this point to the end of the capillary. Record the height to ±1 mm. NOTE: This height will be different from the length of the tube itself. The capillary does not extend all the way down to the end of the sealed glass tubing.

Using tap water at room temperature, repeat the procedure. Enter the temperature and height measurement in the report sheet. Now, fill the graduated cylinder with loosely packed crushed ice. Slowly pour enough water over the ice to fill the cylinder. Immerse the glass tube and thermometer into the ice mixture and take the height and temperature readings. Record the values on the report sheet. Be sure to wait long enough at each temperature.

To obtain a lower temperature bath than 0°C, loosely pack the graduated cylinder with crushed ice again and pour enough methyl alcohol over it to fill the cylinder. Immerse the glass tube and thermometer, and allow the gas to reach a steady temperature. Record the temperature and height on the report sheet.

Set your tube aside while you analyze your data and plot the graph described below. When you find that your results are satisfactory, return the tube or dispose of it as directed by your instructor.

CALCULATIONS

Plot your height versus temperature readings on the graph paper provided and extrapolate the line to V = 0 (corresponding to height = 0). Record this temperature as "absolute zero."

A straight line, such as the one shown in Figure 12-2, is a graphical form of a direct proportionality,

$$V \propto T \qquad (12.4)$$

where V is the volume measured in height and T is the absolute temperature, or number of degrees above your experimental absolute zero. The statement of this proportionality is Charles' Law: at constant pressure, the volume occupied by a fixed quantity of gas is directly proportional to the absolute temperature. Proportionality 12.4 may be converted to an equation by introducing a proportionality constant, k:

$$V = kT \qquad (12.5)$$

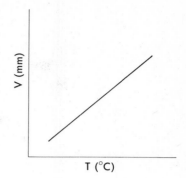

Figure 12-2. Graphical representation of Charles' Law.

The value of the proportionality constant is the slope of the line. Solving Equation 12.5 for the slope, k, we have

$$k = \frac{V}{T}$$ (12.6)

Select two fairly widely separated points *on the line you have drawn* (not points you have plotted) and, from them, calculate the slope. According to the rules of algebra, the equation for slope is

$$slope = \frac{y_2 - y_1}{x_2 - x_1}$$ (12.7)

Record this value on your report sheet, and include its units.

EXPERIMENT 12
REPORT SHEET

Name _____

Date _____ Section _____

Bath	Temperature (°C)	Height of column (mm)
Warm water		
Tap water		
Ice and water		
Ice and methyl alcohol		

Value of "absolute zero" in °C (from graph): _____ .

Slope of the line (k): _____ (mark units).

Name _____

Date _____ Section _____

Height (mm)

Temperature (°C)

Height vs. Temperature

EXPERIMENT 12
ADVANCE STUDY ASSIGNMENT

Name _____

Date _____ Section _____

1. Gas A has particles twice as heavy as those of gas B. If the kinetic energy of both gases is the same, predict which of the particles (A or B) will have the greater velocity. Explain.

2. Why is it important that the atmospheric pressure remain constant during this experiment?

3. Why should you use narrow-bore tubing for this experiment?

Molar Weight of a Volatile Liquid

PERFORMANCE GOALS

13-1 Determine experimentally the mass of a vapor occupying a known volume at a given temperature and pressure.

13-2 Calculate the molar weight of a volatile liquid from the mass of a given volume of vapor at a measured temperature and pressure.

CHEMICAL OVERVIEW

An "ideal gas" is one that behaves as if the molecules (a) are widely separated from each other, (b) have negligible volumes of their own compared to the space they occupy and (c) are not subject to inter-molecular attractive or repulsive forces. Under ordinary conditions in the laboratory, most gases approach the behavior of ideal gases.

In describing the behavior of gases, four variables must be considered: (a) volume, (b) pressure, (c) temperature and (d) amount (number of moles, or mass). Considering the separate relationship between volume and each of the other three variables, we derive these laws:

Volume is inversely proportional to pressure at constant temperature and amount (Boyle's Law):

$$V \propto \frac{1}{P} \qquad (13.1)$$

Volume is directly proportional to absolute temperature (°K) at constant pressure and amount (Charles' Law):

$$V \propto T \qquad (13.2)$$

Volume is directly proportional to the amount of a gas at constant pressure and temperature:

$$V \propto n \qquad (13.3)$$

121

Combining these relationships yields the ideal gas law:

$$PV = nRT \qquad (13.4)$$

where P is pressure, V is volume, n is number of moles of gas, R is a proportionality constant and T is absolute temperature.

The units of R depend upon the units used in measuring the four variables. In working with gases in chemistry, you will measure volume in liters and temperature in degrees Kelvin (converted from degrees Centigrade). If pressure is measured in atmospheres,

$$R = 0.0821 \frac{\text{(liter) (atmospheres)}}{(^{\circ}\text{K) (mole)}} \qquad (13.5)$$

If, on the other hand, pressure is measured in mm Hg,

$$R = 62.4 \frac{\text{(liter) (mm Hg)}}{(^{\circ}\text{K) (mole)}} \qquad (13.6)$$

A useful variation of the ideal gas equation may be derived as follows: If the weight of any chemical species, g, is divided by the molar weight, MW, the quotient is the number of moles.

$$\frac{g}{MW} = \frac{\text{grams}}{\text{grams/mole}} = \text{moles.}$$

Therefore, $\frac{g}{MW}$ may be substituted for its equivalent, n, in Equation 13.4:

$$PV = \frac{g}{MW} RT \qquad (13.7)$$

In this experiment, we shall measure the volume of vapor occupied by a volatile liquid at atmospheric pressure and the temperature of boiling water. Its mass will be determined by condensing the vapor and weighing it. With pressure, volume, temperature and mass all known, molar weight may be determined by direct substitution into Equation 13.7.

SAMPLE CALCULATION

Find the molar weight of a gas if a 0.895 gram sample occupies 235 milliliters at 95°C and 1.02 atmospheres pressure.

Substitution into Equation 13.7 requires that volume be measured in liters and temperature in degrees Kelvin. These conversions are:

$$V = 235 \text{ ml} \times \frac{1 \text{ liter}}{1000 \text{ ml}} = 0.235 \text{ liters}$$

$$°K = °C + 273 = 95 + 273 = 368°K$$

Solving Equation 13.7 for molar weight and substituting the experimental values yields

$$MW = \frac{gRT}{PV} = \frac{0.895 \text{ g}}{1.02 \text{ atm}} \times \frac{0.0821 \text{ (liter) (atm)}}{(°K) \text{ (mole)}} \times \frac{368°K}{0.235 \text{ liter}} = 113 \text{ g/mole}$$

PROCEDURE

Weigh a clean, dry 125 ml Erlenmeyer flask, along with a 2" × 2" square of aluminum foil, on a milligram balance; record the weight in the report sheet. Obtain 4 to 5 ml of an unknown liquid from the instructor and place it in the flask. Record the number of the unknown. Crimp the aluminum foil fairly tightly around the neck of the flask and punch a tiny hole with a pin in the center of the foil.

Fill a 600 ml beaker about 1/3 full of water. Clamp the Erlenmeyer flask to a ringstand and immerse it in the beaker, as in Figure 13-1. Adjust

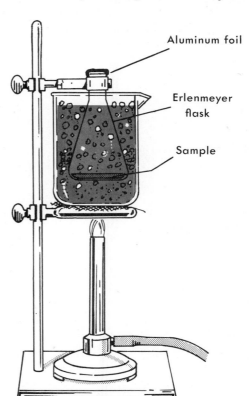

Aluminum foil

Erlenmeyer flask

Sample

Figure 13-1. Apparatus for determining the molar weight of a volatile liquid.

the water level so that the flask is totally immersed, but not so deep that the water reaches the lower edge of the foil.

Remove the flask from the beaker and heat the water to boiling. Return the flask to its former position in the beaker and continue heating. Measure and record the temperature of the boiling water and the barometric pressure. As soon as *all* liquid in the flask has vaporized (no condensed beads or liquid on the walls or neck of the flask), remove the flask from the water and allow it to cool to room temperature. You may find it necessary to remove the flask from the beaker a few times to inspect it for remaining liquid, returning it promptly if liquid is present. As the flask cools, carefully wipe the outside, making sure that no water droplets have collected under the edge of the aluminum foil. Weigh the flask, foil and condensed liquid on a milligram balance and record the weight on your report sheet.

To establish the volume of the gas, fill the flask completely with water and then measure its volume using a graduated cylinder. Record your reading to the nearest milliliter.

CALCULATIONS

From the data, determine the volume of the gas in liters and the vapor temperature in degrees Kelvin. Then find the molar weight, as shown in the sample calculation.

EXPERIMENT 13
REPORT SHEET

Name _____

Date _____ Section _____

Unknown No. _____

DATA

Mass of flask and foil and condensed vapor _____ g

Mass of flask and foil _____ g

Volume of flask _____ ml

Temperature of boiling water _____ °C

Barometric pressure _____ mm Hg

RESULTS

Mass of unknown (condensed vapor) _____ g

Volume of flask (vapor) _____ liters

Temperature of vapor _____ °K

Molar weight of gas _____ g/mole

(Show calculations)

EXPERIMENT 13
ADVANCE STUDY ASSIGNMENT

Name _____

Date _____ Section _____

1. How would each of the following errors affect the outcome of this experiment? Would it make the molar weight high or low? Give your reasoning in three sentences or less in each case.

   a. All of the liquid has not yet vaporized before you remove the flask.

   b. The flask was not completely dried before the final weighing.

2. A volatile liquid was allowed to evaporate in a flask having a total volume of 264 ml. The temperature of the water bath was 100°C at the atmospheric pressure of 0.986 atmospheres. Calculate the mass of the vapor filling the flask if the liquid has a molar weight of 124 g/mole.

Molar Weight Determination By Freezing Point Depression

PERFORMANCE GOALS

14-1 Measure the freezing point of a pure substance.

14-2 Determine the freezing point of a solution by graphical methods.

14-3 Knowing the molal freezing point constant and having determined experimentally the mass of the unknown solute, the mass of a known solvent and the freezing point depression, find the molar weight of the solute.

CHEMICAL OVERVIEW

When a solution is prepared by dissolving a certain amount of solute in a pure solvent, the properties of the solvent are modified by the presence of the solute. The change in such properties as the melting point, boiling point and vapor pressure is found to be dependent on the number of solute molecules in a given amount of solvent. The nature of the solute particles (molecules or ions) is not important; the governing factor is the *relative number of particles*. Properties which are dependent only on the concentration of solute particles are referred to as *colligative properties*.

If a pure liquid is cooled, the temperature will decrease until the freezing point is reached. With continued cooling, the liquid gradually

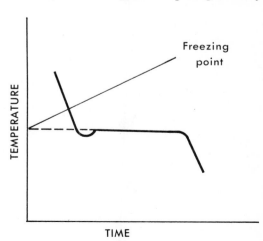

Figure 14-1. Cooling curve of a pure liquid.

freezes. As long as liquid and solid are *both* present, the temperature will remain constant. When all the liquid is converted to solid, the temperature will drop again. A typical cooling curve for a pure solvent is shown in Figure 14-1. The dip below the freezing point is the result of supercooling, an unstable situation in which the temperature drops below the normal freezing point until crystallization begins. Supercooling may or may not occur in any single freezing process and will probably vary in successive freezings of the same sample of pure substance. As soon as crystal formation begins, the temperature rises to the normal freezing point and remains constant until all of the liquid is frozen.

The freezing point of a solution is always lower than that of the pure solvent. The difference between the freezing points of the solvent and solution is the *freezing point depression*. This difference is proportional to the molal concentration of solute. During the freezing of a solution, it is the solvent that freezes. Therefore, the solvent is gradually removed from the solution as the freezing progresses, leaving behind an increasingly concentrated solution. Due to this concentration increase, the freezing point drops, producing a solution freezing curve such as that in Figure 14-2. Again, the supercooling effect may or may not be observed in any single freezing event.

Colligative properties may be utilized to determine the molar weights of unknown substances. It is known, for example, that 1 mole of a nonvolatile nonelectrolyte lowers the freezing point of one kilogram of

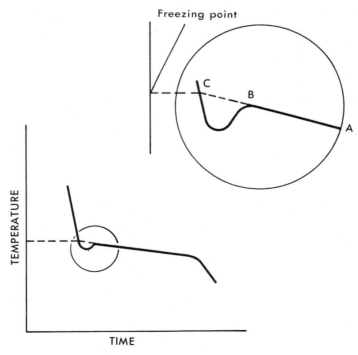

Figure 14-2. Cooling curve of a solution.

water by 1.86°C and elevates its boiling point by 0.52°C. In general, the relationship between freezing point lowering and concentration of solute can be expressed as

$$\Delta T_F = K_F\, m \qquad (14.1)$$

where ΔT_F is the freezing point depression in °C, K_F is the molal freezing point constant of the pure solvent and m is the molality of the solute (moles per kg of solvent).

The solvent you will use in this experiment is naphthalene ($K_F = 6.9°C/m$). You will determine the freezing point experimentally with *your* thermometer. (The literature value of the freezing point is 80.2°C.) You will then prepare a solution from a weighed quantity of naphthalene and a weighed quantity of unknown solute, and determine experimentally its freezing point. When you substitute ΔT_F into Equation 14.1, the molality of the solution can be calculated.

Molality, by definition, is moles of solute per kilogram of solvent. This may be expressed mathematically as

$$m = \frac{\text{mole solute}}{\text{kg solvent}} = \frac{\text{g solute }/MW}{\text{kg solvent}} \qquad (14.2)$$

After you have calculated the molality of the solution prepared with weighed quantities of both solute and solvent, the molar weight (MW) is the only unknown in Equation 14.2.

PROCEDURE

1. THE FREEZING POINT OF NAPHTHALENE

Assemble the apparatus shown in Figure 14-3, using a thermometer graduated in tenths of degrees Celsius. A circular stirrer may be formed from No. 18 wire. A 600 ml beaker half full of water may be used as the water bath. Weigh an empty test tube to ±0.01 g on a centigram balance, add about 10 g of naphthalene and weigh again. Insert the thermometer in the test tube and place the unit in the water bath. Heat the water bath until the naphthalene completely melts. Be sure the entire thermometer bulb is submerged in the molten naphthalene. Discontinue heating, remove the water bath and allow the liquid to cool slowly. Beginning at about 84°C, record the temperature to ±0.1°C every 30 seconds for ten minutes. Stir continuously during this period. After the first crystals appear, the temperature should stay constant until all of the naphthalene freezes. This temperature is the freezing point of pure naphthalene. The theoretical freezing point of naphthalene is 80.2°C but you may obtain a

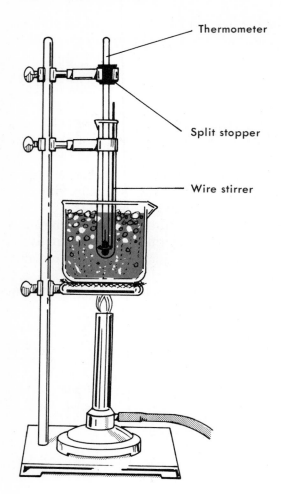

Thermometer

Split stopper

Wire stirrer

Figure 14-3.

slightly different reading, due to error in the calibration of your ther-mometer or small amounts of impurities in the sample. (Question: Do you think such errors will influence the molar weight you are determining?)

2. THE FREEZING POINT OF THE SOLUTION

Weigh out from 0.9 to 1.0 g of unknown solute to an accuracy of ±0.01 g. Carefully transfer the solid into the test tube containing the naphthalene and weigh the test tube and naphthalene to obtain the exact weight of the unknown used. Start heating the mixture in the water bath until all of the naphthalene is melted and the unknown is completely dissolved. With constant stirring, determine the freezing point of the solution exactly the same way as you determined the freezing point of the pure naphthalene.

If time permits, run a second cooling curve determination on your solution.

To clean your equipment, melt the solution in the test tube and pour it onto a paper towel folded to several thicknesses. The test tube may be cleaned with benzene, acetone or some other suitable solvent.

CAUTION: DO NOT POUR MELTED NAPHTHALENE SOLUTIONS INTO THE SINK. THEY FREEZE AND CLOG DRAINS. MANY CLEANING SOLVENTS ARE FLAMMABLE AND SHOULD NOT BE USED NEAR AN OPEN FLAME. PREFERABLY, DO THE CLEANING IN A HOOD.

CALCULATIONS

Determine and record the grams of naphthalene and unknown solute. Estimate to ±0.1°C the freezing point of pure naphthalene from your data. According to Figure 14-1, the freezing point of a pure substance is the constant temperature at which the substance freezes. With due allowance for supercooling, if any, this temperature should be apparent from the table. Record it on the report sheet.

To estimate the freezing temperature of the solution of the unknown in naphthalene, plot the temperature vs. the time on the graph paper provided. The freezing point of the solution *of the concentration you prepared* is the point where it first begins to freeze if there is no super-cooling. This temperature may be estimated from the graph by extrapolating the curve during freezing (from A to B in Figure 14-2) back to its intersection with the cooling line for the liquid (point C). Record the freezing temperature of the solution in the report sheet.

From the freezing points of the naphthalene and the solution, calculate the freezing point depression.

Using Equation 14.1, calculate the molality of your solution.

Using the weights of naphthalene, the unknown solute and the molality previously calculated, substitute into Equation 14.2 and solve it for the molar weight of the unknown. Record it on the report sheet.

EXPERIMENT 14
REPORT SHEET

Name _____

Date _____ Section _____

DATA

Time — Temperature Readings

Time (Minutes)	Naphthalene	Naphthalene + Unknown
0	_____	_____
½	_____	_____
1	_____	_____
1½	_____	_____
2	_____	_____
2½	_____	_____
3	_____	_____
3½	_____	_____
4	_____	_____
4½	_____	_____
5	_____	_____
5½	_____	_____
6	_____	_____
6½	_____	_____
7	_____	_____
7½	_____	_____
8	_____	_____
8½	_____	_____
9	_____	_____
9½	_____	_____
10	_____	_____

Name _____

Date _____ Section _____

Cooling curve of solution

EXPERIMENT 14
REPORT SHEET

Name _____

Date _____   Section _____

Page 2

Unknown No.: _____

Data	Run 1	Run 2
Weight of test tube	g	g
Weight of test tube and naphthalene	g	g
Weight of naphthalene	g	g
Freezing point of naphthalene	°C	°C
Weight of test tube, naphthalene and sample	g	g
Weight of sample	g	g
Freezing point of solution (from curve)	°C	°C
Calculations		
Freezing point depression	°C	°C
Molality	$\dfrac{\text{moles}}{\text{kg}}$	$\dfrac{\text{moles}}{\text{kg}}$
Molar weight of the sample	$\dfrac{\text{g}}{\text{mole}}$	$\dfrac{\text{g}}{\text{mole}}$

EXPERIMENT 14

ADVANCE STUDY ASSIGNMENT

Name _____

Date _____   Section _____

1. Calculate the freezing point of a solution prepared by dissolving 35.0 grams of $C_3H_8O_3$ in 250 grams of water. ($K_F = 1.86°C/m$ for water.)

2. If 1.8 grams of unknown solute are dissolved in 12 grams of solvent, the freezing point depression is 5.0°C.
   a. Calculate the molality of the solution if $K_F = 4.0°C/m$ for the solvent.
   b. Calculate the molar weight of the solute.

The Conductivity of Solutions—A Demonstration

PERFORMANCE GOALS

15-1 Describe how the conductivity of a solution may be tested.

15-2 Basing your decision on conductivity observations, classify substances as strong electrolytes, weak electrolytes or nonelectrolytes.

15-3 Explain the presence or absence of conductivity in an aqueous solution.

CHEMICAL OVERVIEW

Solutions of certain substances are conductors of electricity. The conductance is due to the presence of charged species (ions) which are free to move through the solution. These ions are already present in solid *ionic compounds*. They are simply "released" from the crystal when the compound dissolves, and thereby become mobile. An example is sodium chloride, NaCl, which may be written $Na^+Cl^-(s)$ to emphasize its character as an ionic solid:

$$Na^+Cl^-(s) \xrightarrow{H_2O} Na^+(aq) + Cl^-(aq) \qquad (15.1)$$

With some *molecular compounds,* ions are formed by reaction of the solute with water. For example, hydrogen chloride gas, $HCl(g)$, reacts with water and yields hydrochloric acid:

$$HCl(g) + H_2O(l) \longrightarrow H_3O^+(aq) + Cl^-(aq) \qquad (15.2)$$

In dilute hydrochloric acid, this reaction is virtually complete, and no appreciable number of neutral HCl molecules are present. Solutions that contain a large number of ions are *good conductors*. The solutions, and the solutes that produce them, are called **strong electrolytes.**

When acetic acid is dissolved in water, an equilibrium is reached:

$$CH_3COOH(aq) + H_2O(l) \rightleftharpoons CH_3COO^-(aq) + H_3O^+(aq) \quad (15.3)$$

141

Of the original neutral acetic acid molecules, only a small fraction ionizes. Hence, the resulting solution is a *poor conductor.* Solutes that ionize only slightly and the solutions they yield are classified as **weak electrolytes.**

A third class of solutes includes molecular compounds that do not ionize when dissolved in water. The most minute solute particle remains a neutral molecule. Consequently, such solutions do not conduct an electric current. Solutes whose solutions are *nonconductors* are referred to as **nonelectrolytes.**

Solution conductivity can be detected by a conductivity sensing apparatus such as the one shown in Figure 15-1. The metal strips through which the current enters and leaves the solution are called **electrodes.** If the electrodes are immersed in a solution containing mobile ions, the ions conduct "current," and the bulb lights. If the solution contains no ions, no current will flow, and the bulb does not light. The flow or non-flow of current is therefore a clear indication of the presence or absence of ions. The intensity of light produced (which is proportional to the magnitude of current) is a qualitative measure of the number of ions present.

In this experiment, we will observe the conductivity of several solutions. You will be asked to make classifications, based on your experimental observations, of whether the substances tested are strong electrolytes, weak electrolytes or nonelectrolytes.

PROCEDURE

The instructor will set up a conductivity sensing apparatus such as that in Figure 15-1. This apparatus will be used to test the conductivity of

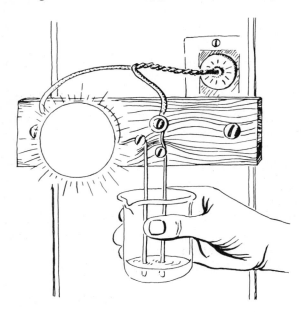

Figure 15-1. Conductivity sensing apparatus.

several solutions. Classify each solution in the following list as a good conductor, a poor conductor or a nonconductor and record your classification in the space provided on the report sheet.

1. Distilled water
2. Tap water
3. Solid NaCl
4. 1.0 M NaCl (salt of strong acid and strong base)
5. Glacial acetic acid, CH_3COOH or $HC_2H_3O_2$
6. 1.0 M $HC_2H_3O_2$
7. 1.0 M HCl
8. 1.0 M NaOH
9. 1.0 M NH_3(aq) ("NH_4OH")
10. 1.0 M $NaC_2H_3O_2$ (salt of weak acid and strong base)
11. 1.0 M $NH_4C_2H_3O_2$ (salt of weak acid and weak base)
12. 1.0 M NH_4Cl (salt of strong acid and weak base)
13. 1.0 M dextrose (sugar), $C_6H_{12}O_6$(180 g/l)

The instructor will mix the solutions in the list that follows. Record any visible evidence of a chemical reaction and note the conductivity of the mixture. Explain your observations in the space provided on the report sheet.

14. Equal volumes of 1.0 M $AgNO_3$ and 1.0 M NaCl
15. Equal volumes of 1.0 M $BaCl_2$ and 1.0 M K_2CrO_4
16. 10 ml 1.0 M $BaCl_2$ and 15 ml Na_2SO_4

EXPERIMENT 15
REPORT SHEET

Name _____

Date _____ Section _____

Substance	Conductor (good, poor, nonconductor)	Electrolyte		
		Strong	Weak	Non-
Distilled Water				
Tap Water				
Solid NaCl				
1.0 M NaCl				
Glacial $HC_2H_3O_2$				
1.0 M $HC_2H_3O_2$				
1.0 M HCl				
1.0 M NaOH				
1.0 M NH_3				
1.0 M $NaC_2H_3O_2$				
1.0 M $NH_4C_2H_3O_2$				
1.0 M NH_4Cl				
1.0 M Dextrose				

1.0 M $AgNO_3$ + 1.0 M NaCl

Evidence of reaction, if any: _____

Did solution conduct electricity? Yes _____ No _____ . Explain.

1.0 M BaCl$_2$ + 1.0 M K$_2$CrO$_4$

Evidence of reaction, if any: _____

Did solution conduct electricity? Yes ____ No ____ . Explain.

1.0 M BaCl$_2$ + 1.0 M Na$_2$SO$_4$

Evidence of reaction, if any: _____

Did solution conduct electricity? Yes ____ No ____ . Explain.

EXPERIMENT 15
ADVANCE STUDY ASSIGNMENT

Name _____

Date _____ Section _____

1. What is a strong electrolyte?

2. How does a conductivity sensing device, such as the one used in this experiment, indicate whether a solution contains ionic or molecular species?

Net Ionic Equations: A Study Assignment

PERFORMANCE GOALS

16-1 Distinguish between an overall equation, a total ionic equation and a net ionic equation.

16-2 Identify spectators in a total ionic equation.

16-3 Given reactants that yield (a) a precipitate, (b) a gas, (c) an un-ionized product or (d) no reaction, write the (1) overall equation, (2) total ionic equation and (3) net ionic equation.

CHEMICAL OVERVIEW

There are several ways to show the changes that occur in a chemical reaction. The most common method is a chemical equation. In a *conventional* equation (also called an *overall* or *molecular* equation), reactant and product compounds are represented by their chemical formulas. The equation is balanced by placing the smallest possible integral coefficients in front of the formulas of substances to make the number of atoms of each element on the left side of the equation equal to the number of atoms of that element on the right side of the equation.

While for some purposes conventional equations are the only satisfactory form for describing reactions, they are clearly inadequate for others. For example, overall equations do not precisely describe the chemical changes that occur at the molecular or ionic level, because the true reactant, the species that actually experiences chemical change, may be an ion which is only a part of the compound from which it is obtained. Actual products may also be ions, which the conventional equation incorporates into formulas of solid compounds which are not, in fact, physically present.

In order to distinguish between substances that are present as ions in solution and those that exist as molecules or as ionic solids, an *ionic* equation is written. A *total ionic equation* includes all chemical species actually present at the scene of a reaction except *solvent* water, which is neither a reactant nor a product, but merely the vehicle in which the reaction occurs. Not all species *present* at the reaction scene undergo chemical change. Substances which are present but experience no chemical change are called *spectators*. A *net ionic equation* is an equation in which the spectators are removed from the total ionic equation. The net ionic equation is limited to those reactants that are actually consumed and those product species that are actually formed in the reaction. In this exercise, we shall direct our attention to writing net ionic equations. Total ionic equations are an intermediate step in writing net ionic equations, and for this reason they may remain unbalanced.

In net ionic equations, it is usually necessary to show clearly the *state* of a reactant or product by writing state designation symbols after the formula of the species in the equation. State designations used include (s) for solids, (l) for liquids, (g) for gases and (aq) for species in aqueous solution.

In order to represent chemical compounds correctly in ionic equations, you must recognize whether a compound will be present in ionic or molecular form. The following criteria may be used.

1. *Strong electrolytes in solution are always present in the form of aqueous ions.* Strong electrolytes include strong acids, strong bases and soluble salts. (See the following explanation for the distinction between strong and weak acids and bases, and between soluble and insoluble salts.)

2. *Weak electrolytes in solution are always present in molecular form.* Weak electrolytes, or un-ionized substances, include weak acids, weak bases and, the most common example of all, water.

3. *Insoluble substances are always present as the total compound; their chemical formulas are written accordingly.* These substances include precipitates and gases.

4. *Unstable substances are written in the form of their decomposition products.* There are three frequently encountered unstable products: sulfurous acid, carbonic acid and "ammonium hydroxide." These species, when formed in a chemical reaction, decompose immediately to the products shown below:

$$H_2SO_3 \ (aq) \longrightarrow H_2O \ (l) + SO_2 \ (g)$$

$$H_2CO_3 \ (aq) \longrightarrow H_2O \ (l) + CO_2 \ (g)$$

$$\text{``}NH_4OH\text{''} \longrightarrow H_2O \ (l) + NH_3 \ (aq)$$

In regard to strong electrolytes, you must be able to distinguish between strong acids and bases, and to identify soluble and insoluble salts. There are only a few strong acids: Hydroiodic, HI; hydrobromic, HBr; hydrochloric, HCl; nitric, HNO_3; and sulfuric, H_2SO_4. Nearly all other dissolved acids, compounds made up of hydrogen ions and any other anion, should be considered weak acids and written in molecular form (e.g., hydrofluoric acid, HF (aq)).

Strong bases are the hydroxides of the alkali metals. These solid substances are readily soluble in water. Barium, strontium and calcium hydroxides are moderately soluble, and should therefore be classified as moderately strong bases. All other hydroxides are insoluble and are shown as solids in ionic equations.

Many salts are water soluble, many are not. Determinations of the solubility of a salt may be made by reference to Table 16-1.

TABLE 16-1. SOLUBILITIES OF IONIC COMPOUNDS*

S — Soluble   I — Insoluble   P — Partly Soluble

Ions	$C_2H_3O_2^-$	F^-	Cl^-	Br^-	I^-	ClO_3^-	OH^-	HCO_3^-	NO_3^-	NO_2^-	SO_4^{2-}	SO_3^{2-}	CO_3^{2-}	S^{2-}	PO_4^{3-}
Li^+	S	P	S	S	S	S	S	S	S	S	S		P	S	I
Na^+	S	S	S	S	S	S	S	S	S	S	S	S	S	S	S
K^+	S	S	S	S	S	S	S	S	S	S	S	S	S	S	S
Ag^+	P	S	I	I	I	S	-		S	P	P	I	I	I	I
NH_4^+	S	S	S	S	S	S	-	S	S	S	S	S	S	S	S
Fe^{2+}	S	P	S	S	S		I		S		S	I	I	I	I
Co^{2+}	S	-	S	S	S	S	I		S		S	I	I	I	I
Ni^{2+}		P	S	S	S	P	I		S		S	I	I	I	I
Mg^{2+}	S	I	S	S	S	S	I		S	S	S	S	I	-	I
Ca^{2+}	S	I	S	S	S	S	P		S	/S	P	I	I	I	I
Ba^{2+}	-	I	S	S	S	S	P		S	S	I	I	I	-	I
Cu^{2+}	P	P	S	-			I		S		S			I	I
Pb^{2+}	S	I	P	P	I	S	I		S	P	I	I	I	I	I
Zn^{2+}	S	P	S	S	S	S	I		S		S	I	I	I	I
Al^{3+}	P	P	S	S	-	S	I		S		S			-	I
Fe^{3+}	-	P	S	S	-		I		S		S			-	I

*A dash (–) indicates an unstable species in aqueous solution; a blank space indicates lack of data.

Since a net ionic equation contains ions which are electrically charged, the total charge on either side of the equation is not necessarily zero. Just as the number of atoms must be the same on both sides of an equation, so also must the algebraic sum of the charges of all ions or molecules (or both) be the same on both sides of an equation. The equation must be balanced both atomically and electrically.

From these principles, we derive a three-step approach to writing net ionic equations.

1. Write the overall equation, showing the formulas of the compounds involved, complete with designations of state. (This equation may be unbalanced.)

2. Write the total ionic equation, in which the dissolved species of the overall equation are shown in the form in which they are actually present — aqueous ions or dissolved molecules. (This equation may be unbalanced.)

3. Write the net ionic equation by eliminating spectator ions from the total ionic equation. (This equation must be balanced.)

EXAMPLES

The following examples are in the form of a program in which each step constitutes a separate "frame."

NOTE: In case you are unfamiliar with this format, you should observe the following directions. First, provide yourself with an opaque shield with which to cover everything beneath the first dotted line that extends across the page. Read to that point. Then, in the spaces provided, write whatever is required. Lower the shield to the next dotted line. The material exposed will be the correct response for the question you have just answered. Compare your own answer to this one, looking back to correct any misunderstanding if the two are different. When you fully understand the first step, read to the next dotted line and proceed as before.

Example 1:
Hydrobromic acid and a solution of ammonium carbonate are combined.

First, write the formula for hydrobromic acid, *including its designation of state.*

_____

- -

1a. HBr (aq).

In inorganic chemistry, the term "acid" nearly always refers to the aqueous solution of the compound. Thus, "hydrobromic acid" and "a solution of hydrogen bromide" are synonymous terms. Had the compound been called simply "hydrogen bromide," the formula would have been HBr (g).

As the first step in the procedure, write the overall equation, including state designations for each reactant and product.

_____

- -

1b. $HBr(aq) + (NH_4)_2CO_3(aq) \rightarrow NH_4Br(aq) + H_2CO_3(aq)$

In writing the overall equation, you must consider the state designations assigned to each product. In finding the ammonium bromide formula, you should have asked yourself, "Will this compound form a precipitate, or will it remain in solution?" This question should lead you to a solubility table such as Table 16-1 or another reference which will confirm that ammonium bromide is, indeed, soluble.

The other product, H_2CO_3, must also be examined. Acid compounds dissolve in water, either as un-ionized products (weak acids) or ionized products (strong acids). At this point, however, the full formula for the acid should be shown, complete with the designation (aq) to show that it is in aqueous solution.

Perhaps, in this example, you did not show H_2CO_3 (aq), but recognized carbonic acid as an unstable compound that decomposes to carbon dioxide and water (see p. 150). If you have not already done so, write the overall equation again, showing the correct end products.

_____

- -

1c. $HBr(aq) + (NH_4)_2CO_3(aq) \rightarrow NH_4Br(aq) + CO_2(g) + H_2O(l)$

Now you are ready for the second step in the procedure, the writing of the overall ionic equation.

_____

- -

1d. $H^+(aq) + Br^-(aq) + NH_4^+(aq) + CO_3{}^{2-}(aq) \rightarrow$
$NH_4^+(aq) + Br^-(aq) + CO_2(g) + H_2O(l)$

The final step in the procedure calls for the net ionic equation to be derived by eliminating the spectator ions, that is, those ions that are unchanged in the reaction and appear on both sides of the equation. Complete the net ionic equation, including balancing.

_____

- -

1e. $2 H^+(aq) + CO_3{}^{2-}(aq) \rightarrow CO_2(g) + H_2O(l)$

The ammonium and bromide ions are spectators in the total ionic equation.

Example 2:
Hydrochloric acid is placed on a marble chip (calcium carbonate). Bubbles appear.

This reaction is quite similar to the previous one, with one significant difference. The difference is in the first step when you write the overall equation, complete with state designation. Write this equation now.

_____

2a. $HCl(aq) + CaCO_3(s) \rightarrow CaCl_2(aq) + CO_2(g) + H_2O(l)$

The difference here is that the wording of the question identifies calcium carbonate as a solid.
Now write the overall ionic equation.

_____

2b. $H^+(aq) + Cl^-(aq) + CaCO_3(s) \rightarrow$

$$Ca^{2+}(aq) + Cl^-(aq) + CO_2(g) + H_2O(l)$$

And finally, write the net ionic equation:

_____

2c. $2 H^+(aq) + CaCO_3(s) \rightarrow Ca^{2+}(aq) + CO_2(g) + H_2O(l)$

Chloride ion is the only spectator to be eliminated.

Example 3:
Solutions of silver nitrate and calcium chloride are combined.

Write the overall equation. Check the solubility of the products.

_____

3a. $AgNO_3(aq) + CaCl_2(aq) \rightarrow AgCl(s) + Ca(NO_3)_2(aq)$

Go on to the overall ionic equation:

_____

_ _

3b. $Ag^+(aq) + NO_3^-(aq) + Ca^{2+}(aq) + Cl^-(aq) \rightarrow$

$$AgCl(s) + Ca^{2+}(aq) + NO_3^-(aq)$$

Elimination of the spectators yields the net ionic equation:

_____

_ _

3c. $Ag^+(aq) + Cl^-(aq) \rightarrow AgCl(s)$

Example 4:
Solutions of sodium hydroxide and hydrochloric
acid are combined.

There is a slight variation in this one, but it should not be trouble-some. Get the overall equation first.

_____

_ _

4a. $NaOH(aq) + HCl(aq) \rightarrow NaCl(aq) + H_2O(l)$

The water produced by the rearrangement of ions is, of course, a liquid, not an aqueous solution.
Now go all the way to the net ionic equation.

_____

_ _

4b. $Na^+(aq) + OH^-(aq) + H^+(aq) + Cl^-(aq) \rightarrow$

$$Na^+(aq) + Cl^-(aq) + H_2O(l)$$

$H^+(aq) + OH^-(aq) \rightarrow H_2O(l)$

This is an example of a reaction going to completion by formation of an un-ionized product, H_2O. Weak acids are also un-ionized, and they are soluble in water.

Example 5:
A solution of potassium acetate is added to nitric acid.

Take it all the way to the net ionic equation.

_____

- -

5a. $KC_2H_3O_2(aq) + HNO_3(aq) \rightarrow KNO_3(aq) + HC_2H_3O_2(aq)$

$K^+(aq) + C_2H_3O_2^-(aq) + H^+(aq) + NO_3^-(aq) \rightarrow$

$$K^+(aq) + NO_3^-(aq) + HC_2H_3O_2(aq)$$

$H^+(aq) + C_2H_3O_2^-(aq) \rightarrow HC_2H_3O_2(aq)$

The net ionic equation is a direct parallel to Example 4, except that the hydroxide ion was replaced by the acetate ion. The spectators, of course, are different.

Example 6:
Solutions of ammonium chloride and sodium hydroxide are combined.

Let's check overall equations this time:

_____

- -

6a. $NH_4Cl(aq) + NaOH(aq) \rightarrow NaCl(aq) + NH_3(aq) + H_2O(l)$

If you wrote NH_4OH as a product, you have forgotten that this compound decomposes to soluble ammonia and water.

The rest of the procedure should follow easily.

6b. $NH_4^+(aq) + Cl^-(aq) + Na^+(aq) + OH^-(aq) \rightarrow$

$$NH_3(aq) + H_2O(l) + Na^+(aq) + Cl^-(aq)$$

$$NH_4^+(aq) + OH^-(aq) \rightarrow NH_3(aq) + H_2O(l)$$

DEMONSTRATION (OPTIONAL)

Observe the following process carried out by the instructor, and write the net ionic equation for the reaction. Also, explain the reason for the changes during the titration.

Procedure. Pour 300 to 350 ml of clear, saturated barium hydroxide solution (carefully decanted from a bottle containing excess solute on the bottom) into a 600 ml beaker. Add 3 drops of phenolphthalein indicator and immerse the electrodes of a conductivity sensing device (see Figure 16-1) into the solution.

In a solution, the current carrying species are ions; hence, good conduction indicates a great number of ions. Conversely, poor conduction shows that a relatively small number of ions are present. Observe whether the solution in the experiment is a good or a poor conductor of electricity.

From a buret, slowly add a 1M sulfuric acid solution to the contents of the beaker. Stir the solution continuously and observe any changes that occur.

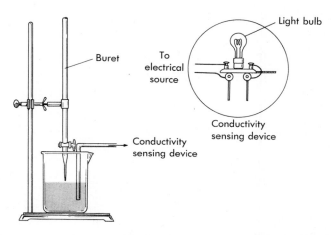

Figure 16-1. Monitoring the conductivity behavior of a solution during a chemical reaction.

EXPERIMENT 16
REPORT SHEET

Name _____

Date _____   Section _____

Write the *net ionic equation* for the following reactions:

1. Sodium hydroxide solution reacts with sulfuric acid.

2. Solid sodium carbonate decomposes when heated, yielding carbon dioxide.

3. Magnesium metal reacts with hydrochloric acid.

4. Barium chloride solution reacts with sulfuric acid.

5. Solutions of calcium chloride and sodium carbonate are combined.

**EXPERIMENT 16
REPORT SHEET**

Name _____

Date _____ Section _____

Page 2

6. A solution of sodium sulfite is added to hydrochloric acid.

7. Solid magnesium carbonate reacts with nitric acid.

8. Solutions of barium chloride and sodium nitrate are combined.

9. Iron (III) nitrate solution is added to sodium hydroxide.

10. Barium hydroxide solution reacts with magnesium sulfate solution.

11. Solid silver acetate reacts with nitric acid.

EXPERIMENT 16
REPORT SHEET

Name _____

Date _____   Section _____

Page 3

12. Sodium hydroxide solution is added to a solution of ammonium nitrate.

13. Acetic acid is neutralized with aqueous ammonia.

14. Ammonium chloride solution reacts with silver nitrate solution.

15. Sodium metal reacts with water, yielding hydrogen gas and a solution of sodium hydroxide.

16. Copper (II) sulfate solution combines with hydrosulfuric acid.

17. Hydrogen chloride is bubbled into water.

EXPERIMENT 16
REPORT SHEET

Name _____

Date _____ Section _____

Page 4

18. Solid sodium oxide reacts with water.

19. Sodium chloride solution combines with phosphoric acid.

20. Solid sodium oxide reacts with sulfuric acid solution.

Demonstration Questions (Optional)

a. Describe and explain what was the color and conductivity of the initial barium hydroxide and phenolphthalein solution?

b. What changes occurred during the titration?

EXPERIMENT 16
REPORT SHEET

Name _____

Date _____   Section _____

Page 5

c. Why did the pink color disappear?

d. Why did the light go out?

e. Why did the light come back on?

f. Write the net ionic equation for the reaction.

EXPERIMENT 16
ADVANCE STUDY ASSIGNMENT

Name _____

Date _____ Section _____

1. What is a spectator?

2. Which of the following substances is *not* ionized in an aqueous solution? Explain.

   a. HCl; b. HNO_3; c. NaCl; d. $HC_2H_3O_2$; e. $BaSO_4$; f. $BaCl_2$

3. Describe the difference between NH_3 (aq) and NH_3 (l).

Titration of Acids and Bases—I

PERFORMANCE GOALS

17-1 Given the volume of a solution of known molarity, and the volume to which it is diluted with water, calculate the molarity of the diluted solution.

17-2 Given the approximate molarity and volume of an acid or base solution to be used in a titration, calculate the number of grams of a known solid base or acid required for the reaction.

17-3 Given the volume of a base or acid solution that reacts with a weighed quantity of a primary standard acid or base, calculate the molarity of the base or acid solution.

17-4 Perform acid-base titrations reproducibly.

CHEMICAL OVERVIEW

Neutralization is the reaction between an acid and a base, yielding water and a salt, usually in solution, as products. Examples of typical neutralization reactions are

$$NaOH(aq) + HCl(aq) \longrightarrow H_2O(l) + NaCl(aq) \qquad (17.1)$$

$$2\,NaOH(aq) + H_2SO_4(aq) \longrightarrow 2\,H_2O(l) + Na_2SO_4(aq) \qquad (17.2)$$

Either the acid or the base in a neutralization reaction may be a solid.

Titration is the controlled addition of a solution into a reaction vessel from a buret. By means of titration, the volume of solution used may be determined quite precisely. The titration process is used in many analytical determinations including those involving acid-base reactions.

An *indicator* is a substance used to indicate when the titration arrives at the point at which the reactants are stoichiometrically, or chemically, equal, as defined by the reaction equation. For example, in an acid-base titration for Equation 17.1, the indicator should tell when the number of moles of NaOH and HCl are exactly equal, matching the 1:1 mole ratio in the equation. For Equation 17.2, the indicator should tell when the moles of NaOH are exactly twice the number of moles of H_2SO_4, this time reflecting the 2:1 molar ratio between the reactants. This point of chemical equality is called the *equivalence point* of the titration.

Acid-base indicators are themselves weak acids or bases that show different colors at different acidities. This color change is due to a structural rearrangement they undergo when the H^+ ion concentration changes. For example, phenolphthalein is colorless in neutral or acidic solutions and pink in basic solutions. Another common indicator, methyl orange, is red in a strongly acidic solution, but yellow in mildly acidic, neutral and all basic solutions. Litmus distinguishes between acids and bases, since its color changes from red in acids to blue in bases. Indicators must be chosen carefully to be sure their color change occurs at the proper point in a specific reaction.

A *standard solution* is a solution with a precisely determined concentration. This concentration is expressed either in molarity or normality. In this experiment, only molarity (moles per liter) will be used. By knowing both the volume and molarity of the standard solution, the number of moles of reagent may be computed:

$$\text{Volume (liters)} \times \text{molarity} \left(\frac{\text{moles}}{\text{liter}}\right) = \text{moles} \qquad (17.3)$$

From the moles of one reactant, moles of the second species are determined according to the equation describing the reaction.

Initially, the concentration of standard solutions is determined from a weighed quantity of a *primary standard*, a highly purified reference chemical. A standard solution may be prepared in either of two ways:

1. A primary standard is carefully weighed, dissolved and diluted accurately to a known volume. Its concentration is calculable from the data.
2. A solution is made to an approximate concentration and then standardized by titrating an accurately weighed quantity of a primary standard.

Once a solution has been standardized in one reaction, it may be used as a standard solution in subsequent titrations. The solution prepared in this experiment will be the standard solution for your next experiment.

In this experiment, you will prepare an NaOH solution of an approximate concentration by diluting a more concentrated solution. From the approximate concentration, you will calculate the amount of oxalic acid, $H_2C_2O_4 \cdot 2H_2O$ (our primary standard), required to react with about 15 ml of the NaOH. That quantity will be weighed into two Erlenmeyer flasks, dissolved and titrated with the NaOH solution, using phenolphthalein as an indicator.

From the weight of oxalic acid, you will be able to determine the number of moles of acid present. From the equation, you will determine the number of moles of NaOH required in the neutralization. You will then know both the volume of the NaOH solution and the number of moles it contains. A calculation of moles divided by volume in liters yields the molarity of the standard solution. The equation for the reaction between oxalic acid and sodium hydroxide is

$$2\,NaOH(aq) + H_2C_2O_4(aq) \longrightarrow 2\,H_2O(l) + Na_2C_2O_4(aq)$$

SAMPLE CALCULATIONS

EXAMPLE 1

25.0 ml of a 12.0 M solution is diluted to 500 ml. Calculate the molarity of the dilute solution.

The number of moles of solute is the same in both the initial solution and the diluted solution; only water is added. This number of moles is

$$0.0250 \text{ liter} \times \frac{12.0 \text{ moles}}{\text{liter}} = 0.300 \text{ mole}$$

In the diluted solution the 0.300 moles solute is dissolved in 500 milliliters, or 0.500 liters. The concentration is therefore

$$\frac{0.300 \text{ mole}}{0.500 \text{ liter}} = 0.600 \frac{\text{mole}}{\text{liter}}$$

EXAMPLE 2

Calculate the molarity of a NaOH solution if a sample of oxalic acid weighing 1.235 grams requires 42.5 ml of the base for neutralization.

First, determine the number of moles of oxalic acid present. The formula of solid oxalic acid is $H_2C_2O_4 \cdot 2H_2O$. Notice that, although the water of hydration is not shown in the equation, the acid is weighed as a solid, which includes the water. Accordingly, calculations must be based on the proper molar weight. Thus,

$$1.235 \text{ g } H_2C_2O_4 \cdot 2H_2O \times \frac{1 \text{ mole } H_2C_2O_4}{126 \text{ g } H_2C_2O_4 \cdot 2H_2O} = 0.00980 \text{ mole}$$

Second, determine the number of moles of NaOH required to react with 0.00980 mole of $H_2C_2O_4$, according to the equation.

$$0.00980 \text{ mole } H_2C_2O_4 \times \frac{2 \text{ moles NaOH}}{1 \text{ mole } H_2C_2O_4} = 0.0196 \text{ mole NaOH}$$

Third, if 0.0196 mole of NaOH is present in 42.5 ml of solution, find the concentration in moles per *liter*.

$$\frac{0.0196 \text{ mole NaOH}}{0.0425 \text{ liter}} = 0.461 \text{ M NaOH}$$

As a single dimensional analysis setup, this calculation would appear as:

$$1.235 \text{ g } H_2C_2O_4\cdot 2H_2O \times \frac{1 \text{ mole } H_2C_2O_4}{126 \text{ g } H_2C_2O_4\cdot 2H_2O} \times \frac{2 \text{ moles NaOH}}{1 \text{ mole } H_2C_2O_4} \times \frac{1}{0.0425 \text{ liter}}$$

$$= 0.461 \text{ M NaOH}$$

PROCEDURE

1. PREPARATION OF NaOH SOLUTION

Obtain 39 to 41 ml of approximately 6 M NaOH and dilute it with 435 to 445 ml of distilled water in a large beaker. Stir thoroughly to assure uniformity and then transfer the solution to a stoppered bottle. Calculate the approximate molarity of the diluted solution.

2. PREPARATION OF OXALIC ACID SOLUTION

Calculate the quantity of oxalic acid needed to neutralize 15 ml of your diluted NaOH solution. Base your calculation on the approximate molarity in the preceding computation. Don't forget the water of crystallization in the oxalic acid and be sure to consider the equation. Have your instructor approve your calculation before proceeding.

Mark and weigh two clean 250 ml Erlenmeyer flasks on a milligram balance. Weigh into each flask an oxalic acid sample that is within 10 per cent of the amount calculated previously. Add about 50 ml distilled water and swirl to dissolve the acid.

3. TITRATION OF OXALIC ACID WITH NaOH

Thoroughly clean a buret. Water should drain off the inner walls without forming droplets. Rinse the wet buret with a 10 ml portion of the dilute NaOH. Drain and repeat with a second portion, discarding it as well. Fill the buret with the NaOH solution and record the initial reading to the nearest 0.1 ml. Make sure that the tip of the buret is filled and that no air bubbles are visible. Place the Erlenmeyer flask under the tip of the buret, add three drops of phenolphthalein solution to the acid sample and start the addition of NaOH. It is essential that you stir the acid solution constantly with a swirling motion and occasionally wash down the walls of the flask. The end point is reached when a light pink color persists for at least 30 seconds. Record the buret reading at the end of the titration.

Repeat the titration with the second sample. You may begin the titration from the point in the buret where the first titration ended, but check and record the reading just prior to titrating. Also, be sure there is sufficient solution in the buret to complete the titration.

The results of this experiment should be calculated immediately. If the molarities calculated by your two titrations are not within 0.006 of each other, a third titration should be run.

CALCULATIONS

Calculate the molarity of your NaOH using the method outlined in Example 2 of the sample calculations. Record these results on your report sheet. As previously instructed, run a third titration if the first two do not agree to within 0.006 M. Label your bottle with the average of your calculated molarities and save it to be used as the standard solution in the next experiment.

EXPERIMENT 17
REPORT SHEET

Name _____

Date _____ Section _____

DATA

	Sample 1	*Sample 2*	*Sample 3*
Mass of flask	_____ g	_____ g	_____ g
Mass of flask + oxalic acid	_____ g	_____ g	_____ f
Mass of oxalic acid	_____ g	_____ g	_____ g
Initial buret reading	_____ ml	_____ ml	_____ ml
Final buret reading	_____ ml	_____ ml	_____ ml
Volume of NaOH solution	_____ ml	_____ ml	_____ ml

CALCULATIONS
(Show setups below.)

Moles of oxalic acid	_____	_____	_____
Moles of NaOH	_____	_____	_____
Molarity of NaOH (moles/l)	_____	_____	_____
Average molarity (moles/l)			_____
Actual molarity of desk reagent (moles/l)			_____

EXPERIMENT 17   Name _____

ADVANCE STUDY ASSIGNMENT

Date _____   Section _____

1. How many milliliters of 12.0 M NaOH would you need to dilute with deionized water to prepare 2.0 liters of 0.500 M solution?

2. A 10.0 ml sample of 0.175 M HCl solution is titrated with an NaOH solution of unknown molarity. If the same NaOH solution is used for the titration of 10.0 ml of 0.125 M H_2SO_4 solution, predict which titration will consume a larger volume of NaOH solution. Justify your prediction.

Titration of Acids and Bases—II

PERFORMANCE GOALS

18-1 Determine the concentration of an acid by titrating a known volume with a standardized base.

CHEMICAL OVERVIEW

In this experiment, you will use the standard NaOH solution prepared in Experiment 17 to determine the molar concentration of either a solution of hydrochloric acid or vinegar, which contains acetic acid ($HC_2H_3O_2$, or CH_3COOH). You will titrate the NaOH into a carefully measured volume of the acid. The product of the volume of NaOH times its molarity is the number of moles of base in the reaction. This product can be converted to the number of moles of acid by means of the reaction equation. The moles of acid divided by the volume containing that number of moles (the volume of your acid sample) yields the molarity of the acid.

PROCEDURE

OPTION 1

Place 225 to 235 ml of water into a 400 ml beaker. Take the beaker, a stirring rod and a 10 or 25 ml graduated cylinder to the hood. Pour 9.5 to 10 ml of concentrated HCl into the graduated cylinder. Slowly, carefully and with constant stirring (CAUTION!) add the acid to the water. Return to your laboratory station and transfer the solution to a labeled bottle.

Clean and rinse a 25.0 ml volumetric pipet and then rinse with your diluted acid solution.

CAUTION! *ALWAYS* USE A RUBBER BULB TO DRAW SOLUTIONS INTO A PIPET. *NEVER* USE MOUTH SUCTION.

Fill the pipet with the acid, adjust the meniscus so that the bottom of it rests on the etched circular mark on the stem and deliver the solution to a

250 ml Erlenmeyer flask. Take two such samples. Add about 50 ml of water (how accurately must this volume be measured?) and three drops of phenolphthalein to each flask.

OPTION 2

Obtain a vinegar solution from your instructor. Pipet 10.0 ml of this solution into each of two 250 ml Erlenmeyer flasks. Add about 50 ml of deionized water and three drops of phenolphthalein to each sample.

Clean and rinse a buret and then rinse it with your NaOH solution from Experiment 17. Fill the buret with NaOH. Carefully read and record the starting volume of your buret. Titrate each acid sample with NaOH, employing the same procedure used in Experiment 17. Your two titrations should include duplicate volumes that differ by no more than 0.3 ml. Be sure you have sufficient NaOH in the buret to complete each titration before you begin.

CALCULATIONS

Calculate the following for each titration and record the results on your report sheet:
1. Approximate concentration of diluted HCl (Option 1 only).
2. Number of moles of NaOH used in titration.
3. Number of moles of acid in 25.0 ml sample (Option 1 only).
4. Molarity of acid.
5. Concentration of original HCl (Option 1 only).
6. Grams of acetic acid in one liter (Option 2 only).

The theory that underlies these calculations is presented in the Chemical Overview.

EXPERIMENT 18
REPORT SHEET

Name _____

Date _____ Section _____

DATA

Option 1

Equation: _____

	Sample 1	Sample 2	Sample 3
Initial buret reading	_____ ml	_____ ml	_____ ml
Final buret reading	_____ ml	_____ ml	_____ ml
Volume of NaOH solution	_____ ml	_____ ml	_____ ml

Molarity of NaOH (from Part I)   _____ M

CALCULATIONS

Approximate concentration of diluted HCl (show calculation):

_____ $\dfrac{\text{moles}}{\text{liter}}$

Number of moles of NaOH used in the titration (show calculation):

_____ moles   _____ moles

Number of moles of HCl in 25.0 ml
diluted acid

_____ moles   _____ moles

Molarity of diluted HCl (show calculation)

_____ $\dfrac{\text{moles}}{\text{liter}}$ _____ $\dfrac{\text{moles}}{\text{liter}}$

Average molarity:

_____ $\dfrac{\text{moles}}{\text{liter}}$

DATA

Option 2

Equation:_____

Initial buret reading   _____ ml   _____ ml

Final buret reading   _____ ml   _____ ml

Volume of NaOH solution   _____ ml   _____ ml

Molarity of NaOH (from Part I)   _____ M

CALCULATIONS

EXPERIMENT 18
REPORT SHEET

Name _____

Date _____   Section _____

Page 2

CALCULATIONS

Number of moles of NaOH used in
the titration (show calculation):

_____ moles   _____ moles

Number of moles of acetic acid in
10.0 ml vinegar

_____ moles   _____ moles

Molarity of acetic acid in vinegar
(show calculation):

_____ $\dfrac{\text{moles}}{\text{liter}}$ _____ $\dfrac{\text{moles}}{\text{liter}}$

Number of grams of acetic acid in 1000
ml solution

_____ g/l   _____ g/l

EXPERIMENT 18   Name _____ _____

ADVANCE STUDY ASSIGNMENT

Date _____   Section _____

1. Calculate the molarity of an acetic acid (CH_3COOH) solution if 50.0 ml of it require 35.8 ml of 0.150 M sodium hydroxide in a titration.

2. Calculate the molarity of the acid obtained when 25.0 ml of 16.0 M reagent is diluted to 750 ml.

3. Calculate the number of milliliters of 0.260 M NaOH required to titrate 25.0 ml of 0.186 M H_2SO_4 solution according to the equation

$$H_2SO_4(aq) + 2\ NaOH(aq) \rightarrow 2\ H_2O(l) + Na_2SO_4(aq)$$

A Study of Reaction Rates

PERFORMANCE GOALS

19-1   State qualitatively and demonstrate experimentally the relationship between the rate of a chemical reaction and (a) temperature, (b) reactant concentration and (c) the presence of a catalyst.

CHEMICAL OVERVIEW

The word "rate" implies change and time — change in some measurable quantity and the interval of time over which the change occurs. Speed in miles per hour, salary in dollars per month or quantity in gallons per minute all express the idea of rate. The rate of a chemical reaction tells us how fast a reactant is being consumed or a product is being produced. Specifically, rate of reaction is the positive quantity that denotes how the concentration of a species in the reaction changes with time. For example, in the reaction

$$N_2 \text{ (g)} + 3 H_2 \text{ (g)} \longrightarrow 2 NH_3 \text{ (g)} \tag{19.1}$$

the rate of reaction can be expressed in terms of NH_3:

$$\text{Rate} = \frac{\text{change in concentration of } NH_3}{\text{time interval}} \tag{19.2}$$

Alternatively, the rate could be expressed in terms of a reactant:

$$\text{Rate} = - \frac{\text{change in concentration of } N_2}{\text{time interval}} \tag{19.3}$$

The minus sign is necessary to make the rate a positive quantity since the concentration of N_2 is decreasing with time.

The rate of a chemical reaction can be changed by (a) varying the concentration of reactants, (b) changing the temperature or (c) introducing a catalyst.* In this experiment, we shall study the effect of concentration by varying the concentration of one reactant while holding

---

*Catalysts are substances that drastically alter reaction rates. These materials are not used up in the reaction and are recovered unchanged at the end of the reaction.

others constant; we shall examine the rates of the same reaction at different temperatures; and we shall conduct two reactions that will be identical in all respects except that a catalyst will be present in one and not in the other.

In analyzing the results of this experiment, *time* of reaction will be used as an indication of rate. Time and rate are inversely related: the higher the rate, the shorter the time. This is evident if you think of driving from one location to another: it takes less time if you drive at a higher rate (speed).

The principal reaction whose rate will be studied in this experiment is the iodine "clock reaction." It is the reaction between two solutions that contain, among other things, an iodide and soluble starch. When the two solutions are combined, a series of reactions begins, ending with the release of elemental iodine. The appearance of iodine in the presence of starch may be detected visually. This signals the completion of the reaction.

PROCEDURE

1. COMPARISON OF REACTION RATES

a. Sodium Oxalate and Potassium Permanganate. Place about 15 ml 0.1 M sodium oxalate into a large test tube and add two droppersful of 6 M sulfuric acid. Place a quantity of water equivalent to the two solutions into a second test tube. This will be used as a control, or a color standard with which to compare the test tube in which the reaction is taking place. Add 8 drops of 0.1 M potassium permanganate to both test tubes and mix by shaking gently or by stirring. Set the test tubes aside in a test tube rack. Record the number of seconds necessary for the reaction to be completed. This is evidenced by the original purple color being replaced by a *colorless* (not tan) solution. Proceed to the next step while you are waiting.

b. The Iodine Clock Reaction. Using your graduated cylinder, pour 10 milliliters of solution A into a test tube. Rinse your graduated cylinder and pour into it 10 milliliters of solution B. Place a small beaker on a piece of white paper. Noting the exact time of mixing, pour the two solutions simultaneously into the beaker. Record the elapsed time before you observe any evidence of a reaction.

2. EFFECT OF CONCENTRATION

To study the effect of concentration on the reaction rate, we shall keep all variables (total volume, concentration of A and temperature) constant except the concentration of solution B. Place four test tubes into a test tube rack and label them 1 to 4. Using a 10 ml graduated cylinder, pour into each test tube the quantities of distilled water and solution B shown in the following table.

Test Tube	Volume of Distilled Water (ml)	Volume of Solution B (ml)
1	2	8
2	4	6
3	6	4
4	8	2

Stir the contents of each test tube with a glass rod. Measure 10 ml of solution A into each of four other test tubes. As in part 1(b), empty the solution from Test Tube No. 1 and one of the solution A test tubes simultaneously into a small beaker on a sheet of white paper, noting the exact time of mixing. Record the number of seconds required for the reaction. Repeat the procedure with the remaining test tube combinations, recording the elapsed time in each case.

3. EFFECT OF TEMPERATURE

Using a graduated cylinder, measure 10 ml of solution A into each of four test tubes. Rinse the cylinder, and then measure 10 ml of solution B into each of four other test tubes. Measure and record the temperature of solution A in one of the test tubes. (We shall assume that solution B is at the same temperature, which is room temperature.) As before, empty one test tube of solution A and one test tube of solution B simultaneously into a small beaker. Record the time required for the reaction.

Place a test tube of solution A and a test tube of solution B into a 250 ml beaker containing tap water and a few chunks of ice. Observe the temperature of solution A. (Again, we shall assume that solution B, treated in an identical manner as solution A, will have the same temperature.) Stir the contents of the beaker with the test tubes. When the temperature drops to about 8 to 10 degrees below room temperature, pour the two solutions into a beaker, as before. Record both the temperature of the solutions and the time of the reaction.

Repeat the above procedure, replacing the ice-water bath with warm tap water. When the temperature of solution A has risen about 10 degrees above room temperature, pour the two solutions together. Obtain one more reading at about 20 degrees above room temperature. For each reading, record the solution temperature and the reaction time in the report sheet.

4. EFFECT OF A CATALYST

Measure 10 ml of solution A into a test tube and add a drop of catalyst. In another test tube, combine 2 ml of solution B and 8 ml of water (the same procedure as in Part 2, Test Tube No. 4). Pour the solutions simultaneously into the beaker and record the time required for the reaction.

Results

Part 2.   Using the graph paper provided, plot a graph of milliliters of solution B over total milliliters versus reaction time.

Part 3.   Using the graph paper provided, plot a graph of temperature versus reaction time.

Part 4.   Compare the reaction times of the catalyzed reaction (Part 4) with the uncatalyzed reaction between solutions of the same concentration (Part 2, Test Tube No. 4). Record your comparison on the report sheet.

EXPERIMENT 19
REPORT SHEET

Name _____

Date _____ Section _____

PART 1 COMPARISON OF REACTION RATES

a. Sodium oxalate + potassium permanganate

   Time of reaction: _____ sec.

b. Iodine clock reaction

   Time of reaction: _____ sec.

PART 2 EFFECT OF CONCENTRATION ON THE IODINE CLOCK REACTION

Tube Number	Volume of Solution B (total volume)	Time (sec)
1		
2		
3		
4		

PART 3 EFFECT OF TEMPERATURE ON THE IODINE CLOCK REACTION

Temperature (°C)	Time (sec)

PART 4 EFFECT OF CATALYST ON THE IODINE CLOCK REACTION

Reaction	Time (sec)
Without catalyst (from Part 2)	
With catalyst	

Which *reaction rate* is greater?

Catalyzed: _____ Uncatalyzed: _____

EXPERIMENT 19
REPORT SHEET

Page 2

Name _____

Date _____ Section _____

RESULTS

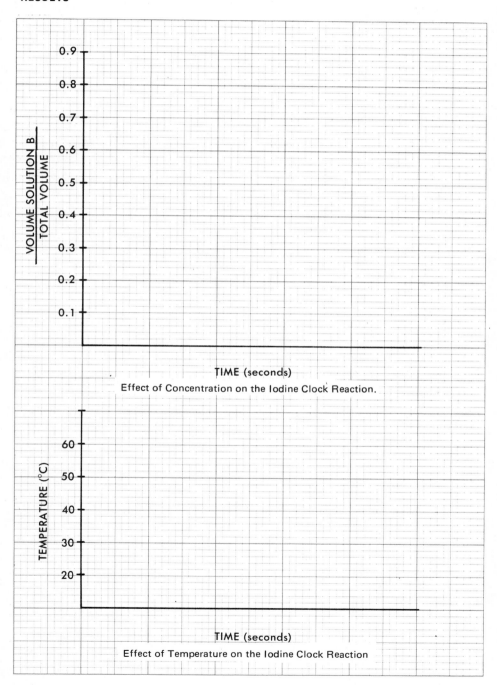

Effect of Concentration on the Iodine Clock Reaction.

Effect of Temperature on the Iodine Clock Reaction

EXPERIMENT 19   Name _____

ADVANCE STUDY ASSIGNMENT

                         Date _____ Section _____

1. Define the meaning of the term "reaction rate."

2. What is a catalyst?

3. If you wanted to alter the rate of a chemical reaction, what changes would you make in the experimental conditions?

Chemical Equilibrium

PERFORMANCE GOALS

20-1 Given the equation for a chemical equilibrium, predict and explain, on the basis of LeChatelier's Principle, the direction of a shift in the position of an equilibrium caused by a change in the concentration of one of the species.

CHEMICAL OVERVIEW

Some chemical reactions *proceed to completion,* that is, until one of the reacting species is for all practical purposes completely consumed. One example of such a reaction is the precipitation of Cl^- ions from solution.

$$Ag^+ (aq) + Cl^- (aq) \longrightarrow AgCl (s) \qquad (20.1)$$

When the Cl^- concentration is essentially zero, the reaction is complete; with no chloride ion available, no more precipitate will form.

Other reactions are *reversible.* This means that when the reactants are introduced into the reaction vessel, the reaction will start, but as soon as the reaction products begin to accumulate, they will react with each other to produce some of the starting species. At any time, *two* reactions are occurring, one going in the *"forward"* direction and one in *"reverse."* For example, consider the dissociation of acetic acid in water:

$$CH_3COOH (aq) \rightleftharpoons CH_3COO^- (aq) + H^+ (aq) \qquad (20.2)$$

While some neutral acetic acid molecules are dissociating, some of the hydrogen ions and acetate ions (CH_3COO^-) formed by the dissociation are recombining to yield the undissociated acid. The two arrows indicate the simultaneous occurrence of two reactions. *When the rate of the forward reaction exactly equals the rate of the reverse reaction, the system is said to be at equilibrium,* and no more visible change occurs. This condition does not mean that all reactions have ceased, but only that the opposing reactions proceed at the same rate.

Consider the generalized reaction

$$A + B \rightleftharpoons C + D \qquad (20.3)$$

199

When the concentration of any one of the species of this equilibrium is changed, the equilibrium is "disturbed," and a "shift" will occur, either in the forward or the reverse direction. *LeChatelier's Principle* predicts the direction of such a shift by stating: *when some stress is applied to a system originally at equilibrium, the system (reaction) will shift in such a direction as to counteract the stress, until a new equilibrium is reached.*

Let us consider how we can apply the preceding principle to the reaction shown in Equation 20.3. Suppose we add more of compound A. What will happen? The outside stress is an increase in the concentration of A. The reaction will shift in a direction that will counteract this increase. That is, the reaction will shift to reduce the concentration of A. A is consumed, and its concentration reduced, if the reaction shifts in the forward direction (as you read the equation from left to right). Similarly, if more C is added, the reverse shift will occur (consuming some of C).

Evidence of a shift in equilibrium can easily be observed in the laboratory if one or more of the substances are colored or if a change in phase, such as precipitation or dissolution, accompanies the shift.

In this experiment, we shall observe qualitatively the effect of changing the concentration of one or more substances in a chemical equilibrium, and we shall correlate these observations with LeChatelier's Principle.

PROCEDURE

1. CHROMATE — DICHROMATE ION EQUILIBRIUM

Study of the chromate–dichromate ion equilibrium is possible because of the different colors of the ions involved:

$$2\,CrO_4{}^{2-}\,(aq) + 2\,H^+\,(aq) \rightleftharpoons Cr_2O_7{}^{2-}\,(aq) + H_2O(l) \qquad (20.4)$$
$$\text{Yellow} \qquad\qquad\qquad\qquad \text{Orange}$$

a. Pour 2 to 3 milliliters of potassium chromate solution into a test tube and add several drops of dilute sulfuric acid. Record in the report sheet any color change observed.

b. Add to the contents of the same test tube several drops of 10 percent sodium hydroxide solution. Shake gently to mix. Record your observations on the report sheet.

Interpret your observations in parts (a) and (b) in terms of a shift of equilibrium in the forward or reverse direction and in terms of LeChatelier's Principle, recording these interpretations in the report sheet.

2. COBALT (II) ION COMPLEXES

Cobalt(II) ions, Co^{2+}, exist in water as aquo-complexes, $Co(H_2O)_6{}^{2+}$, that have a pink color. Other complexes exhibit different colors; the

$CoCl_4{}^{2-}$ complex, for example, is blue. Depending on the relative concentration of chloride ions, the equilibrium shown in the following equation may be altered to yield a solution that is more blue or more pink:

$$Co(H_2O)_6{}^{2+} + 4\,Cl^-\,(aq) \rightleftharpoons CoCl_4{}^{2-} + 6\,H_2O(l) \qquad (20.5)$$
$$\text{Pink} \qquad\qquad\qquad\qquad \text{Blue}$$

To begin this experiment, pour about 2 to 3 milliliters of cobalt chloride solution into each of three test tubes.

a. To the first test tube, add 1 or 2 milliliters of concentrated hydrochloric acid. Note the change, if any. Now add water drop by drop. Is there a change? Note and interpret your observations in the report sheet.

b. To the second test tube, add 1.3 to 1.6 g of solid ammonium chloride and shake to make a saturated solution. Compare the color of the solution with that in the third test tube. Place both tubes in a beaker containing boiling water and note the results. Cool both tubes under tap water. Tabulate and explain your observations.

3. ACETIC ACID DISSOCIATION

Consider the equilibrium

$$CH_3COOH(aq) \rightleftharpoons CH_3COO^-(aq) + H^+(aq) \qquad (20.6)$$

Since no species in this reaction is colored, an auxiliary reagent is needed to help detect any shift in the equilibrium. We will use an indicator, methyl orange, for this purpose, which, in strongly acidic solutions (high H^+ concentrations), is red. A decrease in H^+ concentration will cause a color change from red to yellow and vice-versa.

a. Pour 2 to 3 milliliters of 0.1 M acetic acid into a test tube and add 2 or 3 drops of methyl orange. Place one or two small crystals of sodium acetate in the solution and shake gently to dissolve them. Explain your observations.

b. Repeat the procedure in Part (a) but instead of sodium acetate, add a few drops of 1 M sodium hydroxide. Explain your observations.

4. THE THIOCYANO-IRON(III) COMPLEX ION

This complex ion can be formed from iron(III) ions (Fe^{3+}) and thiocyanate ions (SCN^-) according to the equation

$$Fe^{3+}(aq) + SCN^-(aq) \rightleftharpoons Fe(SCN)^{2+}(aq) \qquad (20.7)$$
$$\text{tan} \qquad\qquad\qquad \text{blood red}$$

Pour 2 or 3 milliliters of 0.1 M iron(III) nitrate and 2 or 3 milliliters of 0.1 M potassium thiocyanate solution into a 100 ml beaker. Dilute with 50 to 55 milliliters of tap water to reduce the intensity of the deep red color. Pour 5 or 6 milliliter portions of this solution into each of three test tubes and proceed as follows.

a.   Add about 1 ml of 0.1 M iron(III) nitrate solution to the contents of the first test tube.

b.   To the second tube, add 1 ml of 0.1 M potassium thiocyanate solution.

c.   To the third test tube, add 6 to 8 drops of 10 percent sodium hydroxide.

Record and explain any changes you observed in parts (a) through (c).

5. SATURATED SODIUM CHLORIDE EQUILIBRIUM

When a saturated solution of sodium chloride is in contact with undissolved solute, the following equilibrium exists:

$$NaCl\ (s) \rightleftharpoons Na^+(aq) + Cl^-(aq) \tag{20.8}$$

Pour 2 or 3 milliliters of saturated solution into a test tube and add a few drops of concentrated hydrochloric acid. Note the result and give an explanation for it.

EXPERIMENT 20
REPORT SHEET

Name _____

Date _____ Section _____

1. Chromate-Dichromate Ion Equilibrium

a. H_2SO_4 *addition:* Color change, if any: _____

Direction of shift (forward, reverse, none): _____

Explanation, according to LeChatelier's Principle:

b. *NaOH addition:* Color change, if any: _____

Direction of shift (forward, reverse, none): _____

Explanation:

2. Cobalt(II) Ion Complex Equilibrium

a. *HCl addition:* Color change, if any: _____

Direction of shift (forward, reverse, none): _____

Explanation:

b. *H_2O addition:* Color change, if any: _____

Direction of shift (forward, reverse, none): _____

Explanation:

c. *NH_4Cl addition:* Colors of solutions in Test Tubes 2 and 3 at different temperatures:

	Test Tube 2	Test Tube 3
Room Temperature		
Boiling Water Temperature		
After Cooling		

Explanation:

EXPERIMENT 20
REPORT SHEET

Name _____

Date _____ Section _____

Page 2

3. **Acetic Acid Dissociation Equilibrium**

   a. *CH_3COONa addition:*   Color change, if any: _____

   Direction of shift (forward, reverse, none): _____

   Explanation:

   b. *NaOH addition:*   Color change, if any: _____

   Direction of shift (forward, reverse, none): _____

   Explanation:

4. **Thiocyano-Iron(III) Complex Ion Equilibrium**

   a. *$Fe(NO_3)_3$ addition:*   Color change, if any: _____

   Direction of shift (forward, reverse, none): _____

   Explanation:

    b. *KSCN addition:*   Color change, if any: _____

        Direction of shift (forward, reverse, none): _____

        Explanation:

    c. *NaOH addition:*   Color change, if any: _____

        Direction of shift (forward, reverse, none): _____

        Explanation:

5. Saturated Sodium Chloride Equilibrium

    Change on adding concentrated HCl: _____

    Explanation:

EXPERIMENT 20   Name _____

ADVANCE STUDY ASSIGNMENT

Date _____ Section _____

1. Show how and why the equilibrium

$$A + B \rightleftharpoons 2\,C$$

will shift if
(a) Extra A is added

(b) Extra B is added

(c) Some C is removed

2. Define, state or describe:
    (a) A reversible reaction

    (b) LeChatelier's Principle

    (c) The characteristics of a system when it is at equilibrium

3. Consider the equilibrium

$$NH_3\,(aq) + H_2O(l) \rightleftharpoons NH_4{}^+\,(aq) + OH^-(aq)$$

Predict the direction the equilibrium will shift upon the
(a) addition of solid NH_4Cl

(b) addition of hydrochloric acid

Measurement of pH with Indicators

PERFORMANCE GOALS

21-1   Prepare a set of pH indicator standards.
21-2   Measure the pH of an unknown solution by using indicators.

CHEMICAL OVERVIEW

Solutions of strong electrolytes such as strong acids and strong bases are good conductors of electricity. This conductivity state is taken to indicate a high concentration of ions. In fact, strong acids and bases break into ions almost completely by either of two processes:

Dissociation is the term used to describe the release of existing ions when an ionic compound dissolves, as in

$$NaOH(s) \longrightarrow Na^+(aq) + OH^-(aq) \qquad (21.1)$$

Ionization is the process whereby ions are formed when a covalent compound reacts with water, as in

$$HCl(g) + H_2O(l) \longrightarrow Cl^-(aq) + H_3O^+(aq) \qquad (21.2)$$

Even though the terms "ionization" and "dissociation" do not mean exactly the same thing, they are closely related and are often used interchangeably.

Solutions of weak electrolytes, such as weak acids and weak bases, by contrast, are poor conductors of electricity. This conductivity state indicates a low concentration of ions. We therefore conclude that weak acids and bases are only partially ionized in water solutions. When acetic acid ionizes by reaction with water, equilibrium is reached:

$$CH_3COOH(aq) + H_2O(l) \rightleftharpoons CH_3COO^-(aq) + H_3O^+(aq) \quad (21.3)$$

At equilibrium, acetic acid is only about 1 percent ionized, compared with HCl, which is nearly 100 percent ionized, as shown by Equation 21.2. Relatively few acetate (CH_3COO^-) and hydronium (H_3O^+) ions are present at equilibrium, but un-ionized acetic acid molecules (CH_3COOH) are in abundance.

The *acidity* of an aqueous solution is a measure of the concentration of the hydrogen (H^+) or hydronium (H_3O^+) ion.

NOTE: The hydronium ion may be considered a hydrated hydrogen ion, $H \cdot H_2O^+$. The H^+ ion is easier to work with and will be used hereafter. It should be understood, however, that this ion is hydrated in aqueous solution and does not exist alone.

A convenient way to express the low acidity of weak acids is to use the **pH** scale. The pH of a solution is mathematically related to the hydrogen ion concentration by the equation

$$pH = -\log [H^+] = \frac{1}{\log [H^+]} \qquad (21.4)$$

where $[H^+]$ is the concentration of the hydrogen ion in moles per liter. By the mathematics of this equation, pH is the negative of the exponent of 10 that expresses the hydrogen ion concentration. For example, if pH = 5, then $[H^+] = 10^{-5}$; and a solution whose pH = 8 has a hydrogen ion concentration of 10^{-8}, or $[H^+] = 10^{-8}$.

Water ionizes into hydrogen and hydroxide ions:

$$H_2O(l) \rightleftharpoons H^+(aq) + OH^-(aq) \qquad (21.5)$$

At 25°C, the ion product of water — the hydrogen ion concentration multiplied by the hydroxide ion concentration — is equal to 1.0×10^{-14}, or

$$[H^+][OH^-] = 1.0 \times 10^{-14} \qquad (21.6)$$

If the ionization of water is the only source of these ions, it follows that they must be equal in concentration:

$$[H^+] = [OH^-] = 1.0 \times 10^{-7} \qquad (21.7)$$

A solution in which the hydrogen ion concentration is equal to the hydroxide ion concentration is said to be **neutral.** The pH of a neutral solution is, by calculation, 7. If the hydrogen ion concentration is greater than the hydroxide ion concentration, the solution is said to be **acidic.** In an acidic solution, the pH is less than 7. Conversely, in **basic** solutions, the concentration of the hydroxide ion exceeds the concentration of the hydrogen ion, and the pH will be greater than 7.

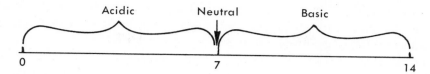

Indicators are organic substances that impart to a solution a color that depends upon its pH. Ordinarily, the color will change gradually over a range of about two pH units. In this experiment, we will use two different indicators in a set of solutions of known pH. By comparing colors, we will then estimate the pH of an unknown solution.

SAMPLE CALCULATIONS

EXAMPLE 1

Calculate the pH of a 0.001 M HNO_3 solution.
Since nitric acid is a strong acid, we assume that it is completely ionized. Therefore,

$$[H^+] = 0.001, \text{ or } 10^{-3} \text{ mole/liter} \qquad pH = 3$$

EXAMPLE 2

Calculate the pH of a solution that contains 0.01 mole of hydrogen ion in 100 ml.
First, calculate $[H^+]$:

$$\frac{0.01 \text{ mole } H^+}{0.100 \text{ liter}} = 0.1 \text{ mole } H^+/\text{liter} = 10^{-1} \text{ mole/liter}$$

It follows that the pH is 1.

PROCEDURE

Since carbon dioxide from the air dissolves in water, yielding an acidic solution, we must remove all dissolved carbon dioxide from the water used in this experiment. Place 350 to 400 milliliters of distilled water in a beaker and heat it to boiling. Continue boiling for approximately ten minutes, cover the vessel and allow it to cool to room temperature.

While the distilled water is being prepared, wash and label six test tubes having a capacity greater than 10 ml. Prepare a set of solutions as follows:

1. Measure into a 50 milliliter graduated cylinder, as accurately as possible, 5.0 milliliters of 1.0 M HCl. Dilute it to 50.0 ml, again very

accurately, with boiled distilled water. Transfer the contents to a small, **dry** beaker and stir thoroughly with a glass rod. Pour 10.0 milliliters of this solution into Test Tube Number 1.

2. Measure 5.0 milliliters of the remaining solution prepared in step (1) into a 50 milliliter graduated cylinder and again dilute carefully to 50.0 ml with boiled distilled water. Transfer again to a clean, **dry** beaker, stir and pour 10.0 milliliters into Test Tube Number 2.

3. Repeat the procedure, using 5.0 milliliters from step (2), diluting to 50.0 milliliters and stirring in a clean, dry beaker. This time, pour 10.0 milliliter samples into *two* test tubes, Number 3 and Number 4.

4. Dilute 5.0 milliliters of the remaining solution from step (3) to 50.0 milliliters in the same fashion and pour 10.0 milliliters into Test Tube Number 5.

5. Finally, dilute 5.0 milliliters from step (4) to 50.0 milliliters and pour 10.0 milliliters into Test Tube Number 6.

Calculate the HCl concentration and, assuming complete ionization, the pH of each of the solutions you prepared. Enter the results in the report sheet.

To each of the test tubes numbered 1 through 3, add three drops of thymol blue indicator and mix well with a glass rod. **Be sure the rod is clean and dry before placing it in each solution.** To test tubes 4 through 6, add two drops of methyl orange indicator and mix in the same fashion. Note the color in each test tube and record your observations in the report sheet. From these observations, estimate the pH range over which each of the preceding indicators changes color.

Obtain one or more unknowns from your instructor. Pour 10.0 milliliters of the unknown into each of two test tubes. To the first test tube, add three drops of thymol blue indicator; to the second, add two drops of methyl orange. Mix thoroughly. Compare the colors of these solutions to your "standard" solutions. Basing your conclusions on the colors you observe, estimate and report the pH of the unknown solution.

EXPERIMENT 21
REPORT SHEET

Name _____

Date _____ Section _____

Test Tube Number	HCl Concentration (moles/liter)	pH	Indicator	Color Observed
1				
2				
3				
4				
5				
6				
Unknown No._____				
Unknown No._____				
Unknown No._____				

Estimated pH range of color transition:

a. Thymol blue: _____

b. Methyl orange: _____

EXPERIMENT 21
ADVANCE STUDY ASSIGNMENT

Name _____

Date _____ Section _____

1. If a solution has a pH of 3.5, is it acidic or basic?

2. Calculate the pH of a 0.001 M HCl solution. If you had an acetic acid solution of the same concentration, would its pH be higher or lower? Explain.

3. 400 milliliters of solution contain 0.00004 mole of HBr. Which of the following indicators would you choose to measure its pH?

Indicator	pH Transition Range
Methyl violet	0.5 to 1.5
Methyl red	4.2 to 6.3
Neutral red	6.8 to 8.0

Explain your reasoning.

Introduction to Oxidation - Reduction Reactions

PERFORMANCE GOALS

22-1 Determine experimentally the relative strengths of a selected group of oxidizing agents.

CHEMICAL OVERVIEW

Oxidation is defined as the process in which a loss of electrons occurs; reduction is a gain of electrons. From a broader viewpoint, in oxidation the oxidation number of an element increases (becomes more positive, as $+3 \rightarrow +5$, or $-3 \rightarrow -1$); whereas in reduction, the oxidation number decreases (becomes more negative, as $0 \rightarrow -1$, or $+7 \rightarrow +2$).

When a metal combines chemically with a halogen to form an ionic compound, an oxidation-reduction (redox) reactions occurs. Electrons are lost by the metal and gained by the halogen. Redox reactions may be thought of as electron transfer reactions, much as acid-base reactions may be viewed as proton transfer reactions. Each redox reaction may be considered the sum of two "half-reactions" or half-cell reactions:

$$M \longrightarrow M^{2+} + 2\ e^-$$ Oxidation of a metal to M^{2+} ion by losing 2 electrons,

and

$$X_2 + 2\ e^- \longrightarrow 2\ X^-$$ Reduction of a halogen to X^- by gaining 1 electron per atom.

Upon addition, we obtain the net ionic redox equation:

$$\begin{aligned}
M &\longrightarrow M^{2+} + 2\ e^- \\
X_2 + 2\ e^- &\longrightarrow 2\ X^- \\
\hline
M + X_2 &\longrightarrow M^{2+} + 2\ X^-
\end{aligned}$$

Observe that the number of electrons lost by the metal exactly equals the number of electrons gained by the halogen. This balance is essential in any redox reaction; there can never be a deficiency or excess of electrons. As an ordinary chemical equation has to be balanced, so does a redox

217

equation. Balancing is achieved by adjusting one or both half-reactions in order to equate the number of electrons lost and gained.

EXAMPLE 1

$$
\begin{aligned}
A &\longrightarrow A + 2\,e^- \quad \text{Oxidation} \\
B + e^- &\longrightarrow B^- \quad\quad\ \text{Reduction}
\end{aligned}
$$

Multiply the reduction half-reaction by 2 and add:

$$
\begin{aligned}
A &\longrightarrow A^{2+} + 2\,e^- \quad \text{Oxidation} \\
2\,B + 2\,e^- &\longrightarrow 2\,B^- \quad\quad\quad\ \text{Reduction} \\
\hline
A + 2\,B &\longrightarrow A^{2+} + 2\,B^- \quad \text{Net ionic redox equation}
\end{aligned}
$$

EXAMPLE 2

$$
\begin{aligned}
A &\longrightarrow A^{3+} + 3\,e^- \quad \text{Oxidation} \\
B + 2\,e^- &\longrightarrow B^{2-} \quad\quad\quad\ \text{Reduction}
\end{aligned}
$$

Multiply the oxidation reaction by 2 and the reduction reaction by 3 to equate the electrons lost in oxidation to the electrons gained in reduction:

$$
\begin{aligned}
2\,A &\longrightarrow 2\,A^{3+} + 6\,e^- \quad \text{Oxidation} \\
3\,B + 6\,e^- &\longrightarrow 3\,B^{2-} \quad\quad\quad\ \text{Reduction} \\
\hline
2\,A + 3\,B &\longrightarrow 2\,A^{3+} + 3\,B^{2-} \quad \text{Net ionic redox equation}
\end{aligned}
$$

In each of the preceding examples, the first element loses electrons to the second element; that is, the first element provides the electrons that reduce the second. Thus, the first element is referred to as a **reducing agent**. By accepting electrons, the second element causes the oxidation of the first element. Hence it is called an **oxidizing agent**. A summary of these terms is presented below:

If a species	the species undergoes	and is called the
Gains electrons	Reduction	Oxidizing agent
Loses electrons	Oxidation	Reducing agent

As acids vary in their strength (the ease with which they release protons), so reducers vary in their strength (the ease with which they release electrons). Similarly, oxidizers have different tendencies to capture electrons, just as bases vary in their attraction for protons. A strong oxidizer (oxidizing agent) has a great affinity for electrons. In this experiment, we shall investigate the relative ease with which certain metals and halides release electrons and shall thereby build a partial qualitative chart of oxidizer strengths.

PROCEDURE

1. METALS

Obtain one strip each of copper, zinc and lead. Clean one side of these strips with emery paper. Lay the strips side by side on a paper towel on the desk, cleaned surface up. Place on each strip one drop of each solution shown in Table 22-1.

TABLE 22–1. SOLUTIONS FOR TESTING METALS

Metal	Test Solutions
Copper, Cu	Zn^{2+}, Pb^{2+}, Ag^+
Zinc, Zn	Cu^{2+}, Pb^{2+}, Ag^+
Lead, Pb	Cu^{2+}, Zn^{2+}, Ag^+

Record in your report sheet the combinations of metal-metal ion that showed evidence of a chemical change and those that did not. Wait about five minutes before you decide that no reaction has occurred. Include the silver ion, Ag^+, although the metal itself is not used because of its high cost. Assume that metallic silver does not react with any of the three solutions.

On completing this part of the experiment, return or dispose of the metal strips as directed by your instructor.

2. HALOGENS

Halogens dissolved in carbon tetrachloride have characteristic colors. Chlorine in carbon tetrachloride is colorless, bromine ranges from tan in dilute solutions to deep red or maroon when concentrated, and iodine is pale pink to deep purple, depending on its concentration. You may observe these colors by placing a drop of bromine water in a test tube containing 3 to 5 drops of carbon tetrachloride, and a drop of iodine solution into another test tube containing CCl_4.

CAUTION: BROMINE IN ELEMENTAL FORM CAUSES SEVERE BURNS ON CONTACT WITH THE SKIN. ALSO, BROMINE AND CHLORINE WATERS RELEASE VAPORS THAT ARE EXTREMELY HARMFUL WHEN INHALED. THESE SOLUTIONS SHOULD BE HANDLED IN A FUME HOOD ONLY, AND WITH UTMOST CARE.

Arrange a series of six of your smallest test tubes in a rack and label them 1 through 6. Into Test Tubes 1 and 2, pour NaCl solution to a depth of about 1/4 inch; into Test Tubes 3 and 4, pour KBr solution to a similar depth; and finally, into Test Tubes 5 and 6, pour KI solution. Add a few drops of bromine water to Test Tube 1 and a few drops of iodine solution to Test Tube 2. Similarly, add a few drops of chlorine water to Test Tube

3, and to Test Tube 4, a few drops of iodine solution. Finally, add a few drops of chlorine water to Test Tube 5 and a few drops of bromine water to Test Tube 6. These different combinations are tabulated as follows:

TABLE 22–2. SOLUTION COMBINATIONS

Test Tube	Halide Solution	Halogen Solution
1	NaCl	Br_2
2	NaCl	I_2
3	KBr	Cl_2
4	KBr	I_2
5	KI	Cl_2
6	KI	Br_2

Now, to each test tube, add carbon tetrachloride to a total depth of about 1/2 inch and shake the test tubes to mix the two solution layers. After shaking, allow the two phases to separate. Water and carbon tetrachloride are immiscible, so carbon tetrachloride, the more dense liquid, will settle to the bottom, carrying any free halogen with it. By the color of this layer, determine which elemental halogen is present after mixing and thus whether or not a redox reaction has taken place.

RESULTS

Record the answers to the following questions in the space provided in the report sheet:

1. Which combination(s), if any, yielded a redox reaction?

2. For each reaction that occurred, write the half reaction equations for both oxidation and reduction.

3. If necessary, multiply either or both equations to equalize the electrons gained and lost in the half-reaction equations.

4. Add the half-reaction equations to get the net ionic redox equation.

5. From Part 1, list the oxidizers in a column according to decreasing strength. Judge oxidizing strength by considering which species each oxidizer was capable of oxidizing and which species it could not oxidize. Show at the bottom the oxidizer that was incapable of oxidizing anything.

6. Prepare a similar list for Part 2.

EXPERIMENT 22
REPORT SHEET

Name _____

Date _____ Section _____

Part 1

Metal	Ion in Solution	Reaction	
		Yes	No
Cu	Zn^{2+}		
	Pb^{2+}		
	Ag^+		
Zn	Cu^{2+}		
	Pb^{2+}		
	Ag^+		
Pb	Cu^{2+}		
	Zn^{2+}		
	Ag^+		
Ag	Cu^{2+}		
	Zn^{2+}		
	Pb^{2+}		

Half-Reactions and Net Ionic Redox Equations

*List of Oxidizers in
Order of Decreasing Strength*

_____

_____

_____

_____

EXPERIMENT 22
REPORT SHEET

Name _____

Date _____ Section _____

Page 2

Part 2

Halide	Halogen	Reaction	
		Yes	No
Cl^-	Br_2		
Cl^-	I_2		
Br^-	Cl_2		
Br^-	I_2		
I^-	Cl_2		
I^-	Br_2		

Half-Reactions and Net Ionic Redox Equations

List of Reducers in
Order of Decreasing Strength

_____

_____

_____

EXPERIMENT 22

ADVANCE STUDY ASSIGNMENT

Name _____

Date _____ Section _____

1. Define the following terms:
   a. Oxidizing agent

   b. Reducer

   c. Redox reaction

2. Combine the following half-reactions to produce a balanced net ionic redox equation:

   a. $A \longrightarrow A^{3+} + 3\,e^-$
      $Z + e^- \longrightarrow Z^-$

   b. $B \longrightarrow B^{2+} + 2\,e^-$
      $Y + 3\,e^- \longrightarrow Y^{3-}$

Preparation of Aspirin

PERFORMANCE GOALS

23-1 Beginning with salicylic acid and acetic anhydride, prepare a sample of aspirin.

CHEMICAL OVERVIEW

Chemically speaking, aspirin is an organic ester. Esters are compounds that are formed when an acid reacts with an alcohol:

Acid	Alcohol	Ester	
			(23.1)

where R_1 and R_2 represent alkyl or aryl groups, such as CH_3-, C_2H_5- or C_6H_5-.

High molecular weight esters such as aspirin are generally insoluble in water and can be separated from a reaction mixture by crystallization. Aspirin can be prepared by the reaction of salicylic acid with acetic acid:

Acetic acid	Salicylic acid	Aspirin	
			(23.2)

As the double arrow indicates, the reaction does not go to completion, but reaches equilibrium.

A better preparative method — the one we will use in this experiment — employs acetic anhydride instead of acetic acid. Acetic anhydride may be considered as the product of a reaction in which two acetic acid molecules combine, with the resulting elimination of a water molecule:

(23.3)

227

The anhydride reacts with salicylic acid to yield the ester (aspirin):

| Acetic anhydride (MW=102) | Salicylic acid (MW=138) | Aspirin (MW=180) | (23.4) |

Excess anhydride reacts with the water produced in the esterification, thereby favoring the forward direction and giving a better yield of the desired product. A catalyst, normally sulfuric or phosphoric acid, is used to speed the reaction.

CAUTION: THE ASPIRIN YOU WILL PREPARE IN THIS EXPERIMENT IS RELATIVELY IMPURE AND SHOULD NOT BE TAKEN INTERNALLY.

PROCEDURE

Preweigh a 50 milliliter Erlenmeyer flask on a decigram balance. Add 1.9 to 2.2 grams of salicylic acid and weigh the flask again. Pour 5.0 to 5.5 milliliters of acetic anhydride into the flask in such a way as to wash down any crystals of salicylic acid that may have adhered to the walls. Add five drops of concentrated phosphoric acid (85 per cent) to serve as a catalyst.

CAUTION: BOTH ACETIC ANHYDRIDE AND PHOSPHORIC ACID ARE REACTIVE CHEMICALS WHICH CAN PRODUCE A SERIOUS BURN ON CONTACT WITH THE SKIN. IN CASE OF CONTACT WITH EITHER, WASH THE SKIN THOROUGHLY WITH SOAP AND WATER. AVOID BREATHING ACETIC ANHYDRIDE VAPORS. WASH ANY SPILLAGE FROM THE DESK TOP.

Clamp the flask in a beaker of water supported on a wire gauze (Figure 23-1). Heat the water to about 75°C, stirring the liquid in the flask occasionally with a stirring rod. Maintain this temperature for about 15 minutes, during which the reaction should be complete. *Cautiously* add 2 milliliters of water to the flask to decompose any excess acetic anhydride. Hot acetic acid vapor will evolve as a result of the decomposition.

When the liquid has stopped giving off vapors, remove the flask from the water bath and add 18 to 20 milliliters of water. Let the flask cool for a few minutes, during which time crystals of aspirin should begin to appear. Put the flask into an ice bath to hasten crystallization and increase the yield of the product. If crystals are slow to appear, it may be helpful to scratch the inside of the flask with a stirring rod.

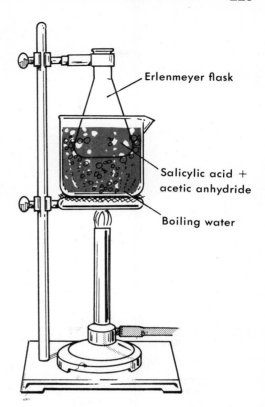

Erlenmeyer flask

Salicylic acid +
acetic anhydride

Boiling water

Figure 23-1. Preparation of aspirin.

Collect the aspirin by filtering the cold liquid through a Büchner funnel, using suction. Turn off the suction, pour about 5 milliliters of ice-cold distilled water over the crystals, and suck down the wash water. Repeat the washing step with a second 5 milliliters of ice-cold water. Draw air through the funnel for a few minutes to help dry the crystals and then transfer them to a clean watch glass that has been preweighed to the nearest 0.1 gram.

In order to determine the yield of aspirin in your experiment, it is necessary that the product be dry. Store the watch glass carefully in your locker. At the beginning of the next laboratory period, weigh the watch glass and aspirin to the nearest 0.1 gram. Record your data in the space provided in the report sheet.

Optional

Very pure aspirin melts at 135°C. By determining the melting point of your aspirin, you may estimate its purity by the closeness of it to 135°. Assemble the apparatus shown in Figure 23-2, using a large oil-filled test tube as the heating bath. Crush some of your aspirin crystals on a watch glass with a spatula. Form a mound from the powder and push the open end of a melting point capillary into the mound. Hold the capillary

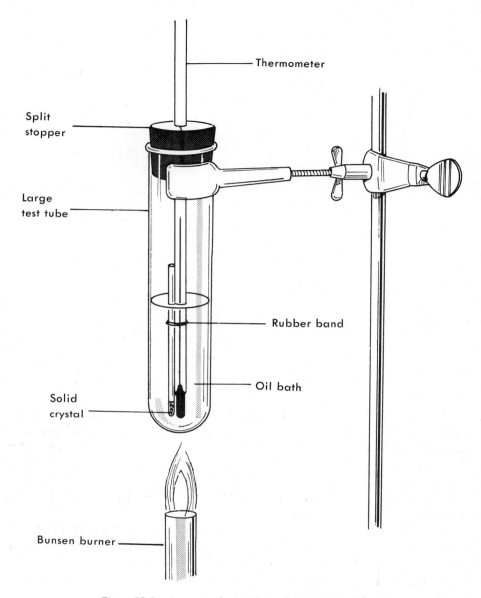

Figure 23-2. Apparatus for melting point determination.

vertically and allow it to drop against the table top, compacting the powder into a plug at the bottom of the tube. Repeating the process, build a plug about 3/4 centimeter to 1 centimeter long. Attach the filled capillary to a thermometer with a rubber band and immerse it in the oil bath. Do not allow the open end of the capillary to come into contact with the oil. Heat the bath rapidly with a Bunsen burner to about 100°C, and more slowly above 100°C. As the melting point is approached, the crystals will begin to soften. Report the melting point as the temperature at which the last crystals disappear (the tube looks transparent).

CALCULATIONS

Based on the actual quantity of salicylic acid used, calculate the theoretical yield of aspirin in grams. Then determine the percentage yield, (actual yield/theoretical yield) X 100. Record your results on the report sheet.

EXPERIMENT 23
REPORT SHEET

Name _____

Date _____ Section _____

DATA AND RESULTS

Weight of 50 ml Erlenmeyer flask   _____ g

Weight of flask plus salicylic acid   _____ g

Weight of salicylic acid   _____ g

Weight of watch glass   _____ g

Weight of watch glass plus aspirin   _____ g

Weight of aspirin (actual yield)   _____ g

Theoretical yield of aspirin, calculated from weight of salicylic acid (show calculations):

_____ g

Percentage yield (show calculations):

_____ %

Melting point of aspirin (optional):   _____ °C

EXPERIMENT 23
ADVANCE STUDY ASSIGNMENT

Name _____

Date _____   Section _____

1. How would you prepare an ester?

2. The name *acetic anhydride* implies that the compound will react with water to form acetic acid. Write the equation for the reaction.

3. Identify, by name or formula, R_1 and R_2 in Equation 21.1 when the ester *aspirin* is formed in this experiment.

Preparation and Properties
of a Soap

PERFORMANCE GOALS

24-1 Starting with a vegetable oil, prepare a soap in the laboratory.
24-2 Examine the chemical properties of the soap you prepared.

CHEMICAL OVERVIEW

An ester is the product of the reaction between an alcohol and a carboxylic acid. The typical equation for the formation of an ester is

$$(24.1)$$

If the alcohol is glycerol, $C_3H_5(OH)_3$, and the acid is a long-chain fatty acid such as stearic acid, $C_{17}H_{35}COOH$, the ester is typical of those found in fats and oils. These esters can be reacted with strong bases to yield glycerol and the salt of the fatty acid. This process is known as *saponification*, and the sodium (or potassium) salts of the fatty acids are called *soaps*. The soap-making process may be written as

$$(24.2)$$

Ester + Base $\longrightarrow$ Glycerol + Soap

where R_1, R_2 and R_3 are long-chain hydrocarbon groups. They may all be the same group, or they may be different groups.

As you can see from Equation 24.2, the anions of soaps contain both polar groups ($-COO^-$) and non-polar groups (long-chain hydrocarbons, R). Polar compounds (or groups) are attracted to water and are called *hydrophilic*. Non-polar compounds (or groups) are water repelling or *hydrophobic* and are soluble in or miscible with non-polar compounds, such as fats, grease, oil or other "dirt." This dual characteristic of soaps is the reason behind their cleaning action. The fat or oil is displaced from the fiber by the soap solution to form large globules that can be detached by jarring (rubbing) and then dispersed (emulsified) in the aqueous solution. *Emulsions* consist of fine droplets of one liquid dispersed in an immiscible liquid (like oil in water).

Generally, soaps made from liquid fats (or oils) are more soluble than those made from solid fats. In the laboratory, we shall prepare a soap by saponifying a vegetable oil with sodium hydroxide. Ethyl alcohol will be added to serve as a common solvent for the reactants. We will also investigate the characteristics of soaps formed from fatty acids and some divalent and trivalent cations. These cations are commonly encountered in areas where there is hard water and in industry. Their soaps are usually referred to as *metallic soaps*.

PROCEDURE

1. PREPARATION OF A SOAP

Weigh a 150 ml beaker on a decigram balance and weigh into it 18 to 20 grams of vegetable oil. Add 20 ml of ethyl alcohol and 25 ml of 20 percent sodium hydroxide solution. Stir the mixture and support the beaker on an asbestos gauze on a tripod. Heat the beaker and its contents gently.

CAUTION: ALCOHOL VAPORS ARE HIGHLY FLAMMABLE. KEEP THE FLAME AWAY FROM THE TOP OF THE BEAKER. HAVE AN ASBESTOS SQUARE HANDY TO COVER THE BEAKER IF THE VAPORS SHOULD IGNITE.

Continue the heating until the odor of alcohol is no longer apparent and a pasty mass remains in the beaker. The reaction product is a mixture of the soap and the glycerol freed in the reaction (see Equation 24.2).

Allow the soap mixture to cool, then add 100 ml of saturated sodium chloride solution and stir thoroughly with a glass rod. This process is called "salting out" and is used to remove the soap from water, glycerol and any excess sodium hydroxide present. After the mixture has been stirred and mixed completely, filter off the soap on a suction funnel (Büchner funnel). Rinse with ice water, drawing this water through the funnel (see Figure 24-1). Allow your soap to dry by spreading it out on a paper towel.

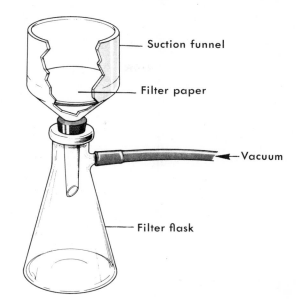

Suction funnel

Filter paper

←Vacuum

Filter flask

Figure 24-1. Vacuum filtration apparatus.

2. PROPERTIES AND REACTIONS OF SOAPS

a. Washing Properties. Take a small amount of your soap and wash your hands with it. In soft water, it should lather easily. If any oil is left over, the soap will feel greasy. Describe the washing properties of the soap on the report sheet.

b. Basicity. A soap that contains free alkali is harmful to skin, silk or wool. To test for the presence of free base, dissolve a small amount of your soap in 5 ml of ethyl alcohol and add two drops of phenolphthalein. If the indicator turns red, free alkali is present. Record your observation.

c. Reaction with Multivalent Cations. Dissolve about 1 g of your soap in 50 ml of warm water. Pour about 10 ml of soap solution into each of three test tubes. To the first test tube, add 8 or 10 drops of 5 percent $CaCl_2$; to the second, 8 or 10 drops of 5 percent $MgCl_2$; and to the third, 8 or 10 drops of 5 percent $FeCl_3$. Record your observations in the report sheet.

d. Emulsification. Put 5 to 10 drops of kerosene in a test tube containing 8 to 10 ml water and shake. An emulsion or suspension of tiny oil droplets in water will form (the solution will look cloudy). Let this solution stand for a few minutes. Prepare another test tube with the same ingredients, but add about 0.5 g of your soap to it before shaking it. Compare the stabilities of the emulsions in the two test tubes. Explain the result.

EXPERIMENT 24
REPORT SHEET

Name _____

Date _____   Section _____

a. **Washing Properties**

b. **Basicity**

c. **Reaction with Multivalent Cations**

Cation Added	Observation
Ca^{2+}	
Mg^{2+}	
Fe^{3+}	

d. **Emulsification**

EXPERIMENT 24
ADVANCE STUDY ASSIGNMENT

Name _____

Date _____ Section _____

1. What is a soap?

2. What does the term saponification mean?

3. What does the term emulsion mean?

4. What do the terms hydrophilic and hydrophobic mean?

APPENDIX

THE OXIDATION NUMBERS OF SOME COMMON CATIONS

+1 *Alkali Metals (Group IA)*	*+2* *Alkaline Earths (Group IIA)*	*+3* *Group IIIA*
Li^+	Be^{2+}	Al^{3+}
Na^+	Mg^{2+}	
K^+	Ca^{2+}	*Transition Elements*
Rb^+	Sr^{2+}	Sc^{3+}
Cs^+	Ba^{2+}	Cr^{3+}
Transition Elements	*Transition Elements*	Fe^{3+}
Cu^+	Mn^{2+}	Co^{3+}
Ag^+	Fe^{2+}	Ni^{3+}
Polyatomic Cations	Co^{2+}	
NH_4^+	Ni^{2+}	
Others	Cu^{2+}	
H^+ or H_3O^+	Zn^{2+}	
	Cd^{2+}	
	Hg^{2+}	
	Hg_2^{2+}	
	Others	
	Sn^{2+}	
	Pb^{2+}	

THE OXIDATION NUMBERS OF SOME COMMON ANIONS

-1	-2	-3
Halogens (Group VIIA)	*Group VIA*	*Group VA*
F^-	O^{2-}	N^{3+}
Cl^-	S^{2-}	
Br^-	O_2^{2-}	*Oxyanions*
I^-	*Oxyanions*	PO_4^{3-}
Oxyanions	CO_3^{2-}	PO_3^{3-}
ClO_4^- BrO_4^- IO_4^-	SO_4^{2-}	BO_3^{3-}
ClO_3^- BrO_3^- IO_3^-	SO_3^{2-}	
ClO_2^- BrO_2^- IO_2^-	CrO_4^{2-}	
ClO^- BrO^- IO^-	$Cr_2O_7^{2-}$	
MnO_4^- NO_3^- $C_2H_3O_2^-$	$C_2O_4^{2-}$	
OH^- NO_2^-	*Acidic anions*	
Acidic anions	HPO_4^{2-}	
HCO_3^- $H_2PO_4^-$	HPO_3^{2-}	
HSO_4^- $H_2PO_3^-$		
HSO_3^-		
Others		
SCN^-		
CN^-		

CONCENTRATIONS OF DESK REAGENTS

Reagent	*Formula*	*Molarity*	*% solute*
Hydrochloric acid, conc.	HCl	12M	37%
Hydrochloric acid, dil.		6	20
Nitric acid, conc.	HNO_3	16	71
Nitric acid, dil.		6	32
Sulfuric acid, conc.	H_2SO_4	18	96
Sulfuric acid, dil.		3	25
Acetic acid, glacial	$HC_2H_3O_2$	17	99.5
Acetic acid, dil.		6	34
Aqueous ammonia, conc.	$NH_3(aq)$	15	29
Aqueous ammonia, dil.	$NH_3(aq)$	6	12
Sodium hydroxide, dil.	$NaOH$	6	20

CHEMICALS REQUIRED FOR THE EXPERIMENTS

Experiment 1

Barium chloride, $BaCl_2$
Sodium sulfate, Na_2SO_4
Barium sulfate, $BaSO_4$
Sodium chloride, NaCl
Barium nitrate, $Ba(NO_3)_2$
Iron(III) chloride, $FeCl_3$
Potassium thiocyanate, KSCN
Sodium carbonate, Na_2CO_3
Copper(II) sulfate, $CuSO_4$
Potassium chromate, K_2CrO_4
Carbon tetrachloride, CCl_4
Methyl alcohol, CH_3OH
Benzene, C_6H_6
3 M Hydrochloric acid, HCl
Conc. Ammonia, $NH_3(aq)$

Note: Arrange solid chemicals into sets, using small vials and a spatula for each bottle.
Anhydrous or hydrated compounds are equally suitable.

Experiment 2

Carbon tetrachloride, CCl_4
Liquid unknown
Solid unknown

Experiment 3

1 M Copper(II) sulfate, $CuSO_4$ (160 g/l)
1 M Iron(III) chloride, $FeCl_3$ (162 g/l)
1 M Nickel(II) sulfate, $NiSO_4$ (155 g/l)
Acetone, CH_3COCH_3
Conc. Hydrochloric acid, HCl
Conc. Ammonia, $NH_3(aq)$
1% dimethylglyoxime in alcohol
Unknown mixtures

Experiment 4

Naphthalene, $C_{10}H_8$
α-naphthol
p-dichlorobenzene
Glacial acetic acid, $HC_2H_3O_2$

Experiment 5

Option 1

Copper wire coil No. 28 (or medium shavings)
Sulfur (flower, powder)

Option 2

Tin foil
Conc. Nitric acid, HNO_3

Option 3

Magnesium ribbon

Experiment 6

Cobalt chloride hexahydrate, $CoCl_2 \cdot 6H_2O$
Sodium sulfate decahydrate, $Na_2SO_4 \cdot 10H_2O$
Calcium chloride (anhydrous), $CaCl_2$
Magnesium sulfate heptahydrate, $MgSO_4 \cdot 7H_2O$
Sodium carbonate decahydrate, $Na_2CO_3 \cdot 10H_2O$
Hydrate unknowns

Experiment 7

Copper shot
Unknown metal

Experiment 8

Potassium chlorate, $KClO_3$
Manganese dioxide, MnO_2

Experiment 10

1 M Sodium carbonate, $Na_2CO_3 \cdot 10 H_2O$ (286 g/l)
1 M Barium chloride, $BaCl_2 \cdot 2 H_2O$ (244 g/l)
0.5 M Sodium sulfate, $Na_2SO_4 \cdot 10 H_2O$ (161 g/l)
0.5 M Potassium chromate, K_2CrO_4 (97.1 g/l)
3% Hydrogen peroxide, H_2O_2
0.5 M Potassium thiocyanate, KSCN (48.5 g/l)

0.5 M Sodium chloride, NaCl (29.2 g/l)
0.5 M Sodium acetate, $Na_2C_2H_3O_2 \cdot 3 H_2O$ (68.0 g/l)
0.5 M Ammonium chloride, NH_4Cl (26.8 g/l)
0.5 M Copper (II) sulfate, $CuSO_4 \cdot 5 H_2O$ (125 g/l)
0.5 M Nickel (II) sulfate, $NiSO_4 \cdot 6 H_2O$ (131 g/l)
0.1 M Iron (III) nitrate, $Fe(NO_3)_3 \cdot 6 H_2O$ (35.0 g/l)
0.1 M Silver nitrate, $AgNO_3$ (17.0 g/l)
1% Dimethylglyoxime in alcohol
1 M Ammonia, NH_3(aq) (67 ml conc. diluted to 1 l)
6 M Hydrochloric acid, HCl
6 M Nitric acid, HNO_3
6 M Acetic acid, $HC_2H_3O_2$
6 M Sodium hydroxide, NaOH (240 g/l)
Conc. Ammonia, NH_3(aq)

Experiment 12

Mercury
Methyl alcohol, CH_3OH

Experiment 13

Aluminum foil, 2" × 2"
Unknowns

Experiment 14

Naphthalene, $C_{10}H_8$
Unknowns

Experiment 15

Sodium chloride, NaCl
Glacial acetic acid, $HC_2H_3O_2$

The following solutions are all 1 M:
Sodium chloride, NaCl (58.5 g/l)
Acetic acid, $HC_2H_3O_2$ (59 ml glacial acetic acid diluted to 1 l)
Hydrochloric acid, HCl (83.5 ml conc. diluted to 1 l)
Sodium hydroxide, NaOH (40.0 g/l)
Ammonia, NH_3(aq) (67 ml conc. diluted to 1 l)

Sodium acetate, $NaC_2H_3O_2$ (82.0 g/l)
Ammonium acetate, $NH_4C_2H_3O_2$ (77.1 g/l)
Ammonium chloride, NH_4Cl (53.5 g/l)
Silver nitrate, $AgNO_3$ (170 g/l)
Barium chloride, $BaCl_2 \cdot 2 H_2O$ (244 g/l)
Potassium chromate, K_2CrO_4 (194 g/l)
Sodium sulfate, Na_2SO_4 (142 g/l)
Sugar, $C_6H_{12}O_6$ (180 g/l)

Experiment 17

Oxalic acid dihydrate, $H_2C_2O_4 \cdot 2H_2O$
6 M Sodium hydroxide, NaOH (240 g/l)
Phenolphthalein solution

Experiment 18

Option 1

Conc. Hydrochloric acid, HCl
Phenolphthalein solution

Option 2

Vinegar solution
Phenolphthalein solution

Experiment 19

0.1 M Sodium oxalate, $Na_2C_2O_4$ (13.4 g/l)
0.1 M Potassium permanganate, $KMnO_4$ (15.8 g/l)
0.1 M Sodium iodide, NaI (15.0 g/l)
Solution A: 50 g KI, 90 mg $Na_2S_2O_3$, 10 ml of 5% soluble starch solution per liter
Solution B: 5 g $Na_2S_2O_8$ per liter
Catalyst: mix equal volumes of 0.1 M copper(II) sulfate, $CuSO_4$ (16.0 g/l) and 0.1 M iron (II) sulfate, $FeSO_4$ (15.2 g/l)
6 M Sulfuric acid, H_2SO_4

Experiment 20

0.1 M Potassium chromate, K_2CrO_4 (19.4 g/l)
0.1 M Cobalt chloride, $CoCl_2 \cdot 6 H_2O$ (24.0 g/l)

0.1 M Acetic acid, $HC_2H_3O_2$ (5.9 ml glacial acetic acid diluted to 1 l)
1 M Sodium hydroxide, NaOH (40.0 g/l)
0.1 M Iron(III) nitrate, $Fe(NO_3)_3 \cdot 6H_2O$ (35.0 g/l)
0.1 M Potassium thiocyanate, KSCN (9.7 g/l)
10% Sodium hydroxide, NaOH (100 g/l)
3 M Sulfuric acid, H_2SO_4
Ammonium chloride, NH_4Cl
Sodium acetate, $NaC_2H_3O_2$
Saturated sodium chloride, NaCl (approx. 380 g/l)
Conc. Hydrochloric acid, HCl
Methyl orange solution

Experiment 21

1 M Hydrochloric acid, HCl (83.5 ml conc. diluted to 1 l)
Thymol blue indicator solution
Methyl orange solution
Unknown acid solutions

Experiment 22

Copper metal strips (about 0.5 cm × 3 cm)
Zinc metal strips (about 0.5 cm × 3 cm)
Lead metal strips (about 0.5 cm × 3 cm)
0.5 M Zinc sulfate, $ZnSO_4 \cdot 7H_2O$ (144 g/l)
0.5 M Lead(II) chloride, $PbCl_2$ (139 g/l)
0.5 M Silver nitrate, $AgNO_3$ (85.0 g/l)
0.5 M Copper sulfate, $CuSO_4 \cdot 5H_2O$ (125 g/l)
0.5 M Sodium chloride, NaCl (29.2 g/l)
0.5 M Potassium bromide, KBr (59.5 g/l)
0.5 M Potassium iodide, KI (83.0 g/l)
Cl_2 water (saturated)
Br_2 water (saturated)
0.05 M Iodine, I_2 (12.7 g/l in methyl alcohol)
Carbon tetrachloride, CCl_4

Experiment 23

Salicylic acid
Acetic anhydride
Conc. Phosphoric acid, H_3PO_4 (85%)
Oil for heating bath

Experiment 24

Vegetable oil
Ethyl alcohol, C_2H_5OH
20% Sodium hydroxide, NaOH (200 g/l)
Saturated sodium chloride, NaCl (approx. 380 g/l)
5% Calcium chloride, $CaCl_2$ (50.0 g/l)
5% Magnesium chloride, $MgCl_2$ (50.0 g/l)
5% Iron(III) chloride, $FeCl_3$ (50.0 g/l)
Phenolphthalein solution
Kerosene

CREDITS FOR ILLUSTRATIONS

Figure 2-1. From Cherim: *Preliminary College Chemistry*. Philadelphia, W. B. Saunders Company, 1973.

Figure 3-1. Adapted from D'Auria, Gilchrist and Johnstone: *Chemistry and the Environment: A Laboratory Experience*. Philadelphia, W. B. Saunders Company, 1973.

Figure 3-2. *Ibid.*

Figure 3-3. *Ibid.*

Figure 4-2. Adapted from Jones and Dawson: *Laboratory Manual for Chemistry, Man and Society*. Philadelphia, W. B. Saunders Company, 1972.

Figure 5-1. *Ibid.*

Figure 12-1. *Ibid.*

Figure 15-1. From Lee, Van Orden and Ragsdale: *Laboratory Manual for General and Organic Chemistry*. Philadelphia, W. B. Saunders Company, 1973.

Figure 23-1. From Jones and Dawson: *Laboratory Manual for Chemistry, Man and Society*. Philadelphia, W. B. Saunders Company, 1972.

Figure 23-2. Adapted from Slowinski, Masterton and Wolsey: *Chemical Principles in the Laboratory*, 2nd edition. Philadelphia, W. B. Saunders Company, 1973.

Figure 24-1. Adapted from Jones and Dawson: *Laboratory Manual for Chemistry, Man and Society*. Philadelphia, W. B. Saunders Company, 1972.